Biotechnology and Molecular Biology

A Laboratory Manual

The Author

Dr. Neelam Yadav is working in the School of Life Sciences, Department of Zoology, Babasaheb Bhimrao Ambedker University, Lucknow. She completed her post graduation in Zoology from Bundelkhand University Jhansi and was awarded Ph.D in 2015 by Bundelkhand University Jhansi. Dr. Neelam has undergraduate teaching experience of 3 years and postgraduate teaching experience of 4 years. She has 8 years of research experience. She has published 07 research papers in reputed national and international journal, one chapter in book.

Biotechnology and Molecular Biology

A Laboratory Manual

– Author –

Dr. Neelam Yadav

School of Life Sciences
Department of Zoology
Babasaheb Bhimrao Ambedker University
Lucknow, Uttar Pradesh

BIOTECH

2026

BIOTECH BOOKS®

ISBN 978-81-7622-503-8

Published by: **BIOTECH BOOKS®**
4762-63/23, Ansari Road, Darya Ganj,
New Delhi - 110 002
Phone: 09899295259, 011-43003222
E-mail: biotechbooks@yahoo.co.in
Website: www.biotechbooks.in

Typeset at: **Classic Computer Services**
Delhi - 110 035

Printed at: **Replika Press Pvt. Ltd.**

PRINTED IN INDIA

Acknowledgement

Many thanks to all whose work, research and support helped me to write this book. I humbly remember and thank my teachers who have provided the basic knowledge and guidance throughout the course of my studies.

I would like to thank Dr. Suman Mishra (Associate professor) and Prof. Kamal Jaiswal (Dean and head) , School of life sciences, Department of Zoology , Babasaheb Bhimrao Ambedker University Lucknow, who have supported and encouraged me for writing this book.

I convey my sincere gratitude to Prof. U.S. Gautam (Hon'ble Vice Chancellor). Banda University of Agriculture and Technology, Banda, Uttar Pradesh (India).

I may Dedicate this book to my parents for their unconditional love, blessings, support and everlasting inspiration.

Dr. Neelam Yadav

Preface

Science can be defined in a variety of ways. Science may be the construction of rules and relationship that define a world that matches our world as closely as possible. The two main branches Biotechnology and Molecular biology are developing branches of science. Which make use of all basic science for industrial, agriculture and medical level applications. In this laboratory manual almost all areas of basic Biotechnology and molecular biology are included. It has various techniques with experimental protocols. Protocols are given in simplified manner and easy to perform. General information about a Biotechnology as well as molecular laboratory is available in this book. Hope this will serve as first hand laboratory manual for graduate, postgraduate and research students.

Dr. Neelam Yadav

Contents

Organizers

Course Faculty and Speakers

List of Participants

Lab Schedule

— Section A —

Molecular Biology

1

Media Preparation

AIM: Preparation of Liquid and Solid Bacterial Culture Medium

Principle

There are various types of microorganisms present in the environment. It is possible to grow the microorganism in laboratory, for their growth and survival in laboratory we can use various types of culture media depending upon the nature of microorganisms. Generally media are classified according to their composition and requirement of organisms.

The medium must contain all nutrients according to requirement which provide the energy, carbon, nitrogen, Phosphorus, sulpher and various minerals *etc.* Nutritional requirement varies from organism to organism. Nutritional requirement reflects the habitat and the surrounding of an organism. Culture media can be classified on the basis of several parameters like chemical composition requirements of an organism:

A. Classification of Media According to their Composition

a. Defined Media or Synthetic Media

When all the components of media are known such type of media is called defined media.

Photolithotrophic autotrops such as cynobacteria, photosynthetic protest, chemoorganotrophic heterotrophs can also can be grown in such type of media.

Composition

Co_2 as carbon sources, Sodium carbonate and sodium bicarbonate can also be added as carbon sources. Nitrate and ammonia are used as nitrogen sources, Sulphate, phosphate and a variety of minerals also can be added in defined media.

Such type of media used in extraction of metabolites and metabolic pathway of experimental microorganisms.

b. Complex Media

If nutritional requirements of an organism are unknown then complex media should be provided because the defined media cannot be constructed. Blood and serum can also be added in such type of medium.

Composition

- ☆ **Peptones:** Peptones are protein hydrolysates prepared by partial photolytic digestion of meat, casein, Soya meal, Geletin and other protein. It always serves as sources of carbon, energy and nitrogen.
- ☆ **Beef Extract:** It is an extract product of a lean beef and is an excellent source of Amino acid, peptides, nucleotides, organic acids, Vitamins and minerals *etc.*
- ☆ **Yeast Extract:** It an extract of Brewer's yeast and provide various types of Vitamins specially Vitamin B, nitrogen and various types of Carbon sources.

Some main complex media which are used in microbiology labs for various researches are:

a) Nutrient broth
b) Tryptic soy broth
c) MacConkey agar

Note

1. Amount or quantity of agar used in solidifying the defined and complex media is approximately 1.0 to 2.0 gram or 1.5 per cent.
2. Agar is a most using solidifying agent in microbial labs because most of the microorganisms cannot be degrade it. There some other solidifying agents besides the Agar can also be used like Silica gel for Autotrophic bacteria.

B. Classification of Media According to Requirements of Microorganisms

a. **Basal Media:** it is required for growth of wide variety of microorganisms. it is composed of very rich components which provide essential supplements' and support the growth of microorganisms.eg: nutrient Agar.

b. **Enrich Media:** It is the media which is used for growth of particular or certain microorganisms. It is enriched with more supplements' according to requirements of microorganisms.

c. **Differential Media:** Such type of media is used in Identification of a specific type of organisms *e.g.*, TSI agar media is used in identification of gram negative bacteria.

d. **Selective Media:** Such type of media is prepared by using various chemicals , dyes and antibiotics, for examples: The combination is prepared with Mac Conkey Agar, Bile salt and crystal violet used as a growth inhibitor for gram positive bacteria and growth stimulus for Gram negative bacteria.

The medium which is prepared with Agar is called **solid medium** and the media which is prepared without Agar is called **liquid media or broth**. For the solidification of media we generally use Agar which Solidifies just below 45°C and liquefies at 90°C. Agar is a Phycocolloid, water soluble polysaccharide. It is a mixture of two valuable polysaccharides. Agarose (Neutral Polymer) and Agro pectin (Acidic Polymer). Agar is obtained from Sea weeds or marine algae. Gelidium, garcilaria, agrophytes (A red sea weed), petrocladia *etc.* are used as good source of agar. It is complex carbohydrate which is mainly composed of D- galactose and 3,6-anhydro-L-galactose and sulphate as polymer.

Requirements

(a) **Glassware and other materials:** Conical flasks, Petri plates, spirit lamp, culture tubes (50 ml), sterile micro pipette-tips, glass pipettes (5 and 10 ml), cotton, gauge cloth *etc.*

(b) **Instruments:** Autoclave, incubator

(c) **Reagents:**

1. Luria-Bertani (LB) medium
2. Agar-agar
3. Antibiotic (such as ampicillin)

Procedure

Preparation of LB Broth/Agar

1. Weigh the appropriate amount of the ingredients in required volume of water and dissolve the components.
2. Adjust the pH and volume to the final and autoclave at 15 lb/in2 for 15 min.
3. Cool the medium to room temperature before use.

4. For agar medium, allow the medium to cool to 45°C (*i.e.*, the temperature of the flask should be bearable when touched by hand) and add appropriate amount of antibiotic (for selection) and dispense into petri plates or slants. Seal with Parafilm and wrap in aluminium foil and store at 4°C till further use

Chemicals

1. Luria-Bertani (LB) Medium

Bacto yeast extract - 5 g

Bacto-tryptone - 10 g

Sodium chloride - 10 g

Dissolve in 900 ml of dH20 and check pH with pH paper, it should be around 7.0. If not, adjust pH with 1 N NaOH to 7.0 and makeup the volume to 1 litre.

2. LB Agar Medium

Bacto yeast extract - 5 g

Bacto-tryptone - 10 g

Sodium chloride - 10 g

Dissolve in 900 ml of dH_2O and check pH with pH paper. it should be around 7.0. if not adjust pH with 1 N NaOH to 7.0. Add

Agar – agar - 15g

Makeup the volume to 1 litre.

2

Isolation of Pure Culture

AIM: To Isolate Pure Colonies of Bacteria from Soil

Principle

There are several types of bacteria present in the environment that grow in complex. In such type of complex, various species present, which interact with other due to which researchers face many problems because it is not possible to determine which organism is responsible for a particular observed activity or disease. In a pure culture, arising from a single cell is characterized by a particular individual species of bacteria. Apure culture is very important in microbiology labs because it requirs study of the concept of growth, characteristics, pathogenecity, metabolism and antibiotics reactivity. Various types of methods which are used for the isolation of pure culture:

a. The streak plate method
b. The spread plate method
c. The pour plate method

a. Streak Plate Method

The streak plate method is done with the help of inoculating loop or swab. In this method a portion or small amount of mixture is streaked out over the surface of the sterile medium in petri-plate. We can repeats it several pattern, if streaking is formed properly then individual colloney or bacteria may be separated from each other.

b. Spread Plate Method

- In such type of method diluted mixture of cells is spread over the agar surface with the help of a **blend glass rod.** This blend rod is called spreader. Diluted microbial mixture contains about 30 to 300 cells. The dispersed cells develop in to isolated colonies, these colonies develop from single cells.

c. The Pour Method

To obtain the separated individual colonies of microorganism the dilution of the sample is increase several times and then mixed with liquid agar medium that has been cooled to 45°C and this mixture is poured immediately in the petri- plate. In such type of method each viable microorganism will developed in to a colony. This method is used to determine the number of cells present in a particular population.

Total number of colonies present in culture = The number of viable microorganisms present in the sample

3

Bacterial Growth Curve

Requirements

(a) Glassware and other materials: Spirit lamp , sterile toothpicks, sterile microfuge tubes. micropipettes (P-20. P-200). sterile micro pipette-tips. conical flasks, culture tubes, cotton, muslin cloth, conical flasks.

(b) Instruments: Shaker incubator, incubator

(c) *E. coli* strain (*e.g.* DH5a)

(d) Reagents: LB medium (broth/agar)

Procedure

A. Inoculation of the Culture (Broth)

1. Take 5 ml of LB medium in a 25 or 50 mi conical flask or culture tube (24 tubes or flasks per experiment).
2. Inoculate with a single colony from a plate.
3. Let the flasks or tubes _grow on shaker incubator at 37°C.
4. Measure the optical density (0.D.) at A550 at intervals of one hour each starting from the zero hour against a blank for 24 hours.

B. Plotting the Graph

1. Plot the readings of O.D. on Y-axis against the time interval on X-axis, which will be a sigmoid.

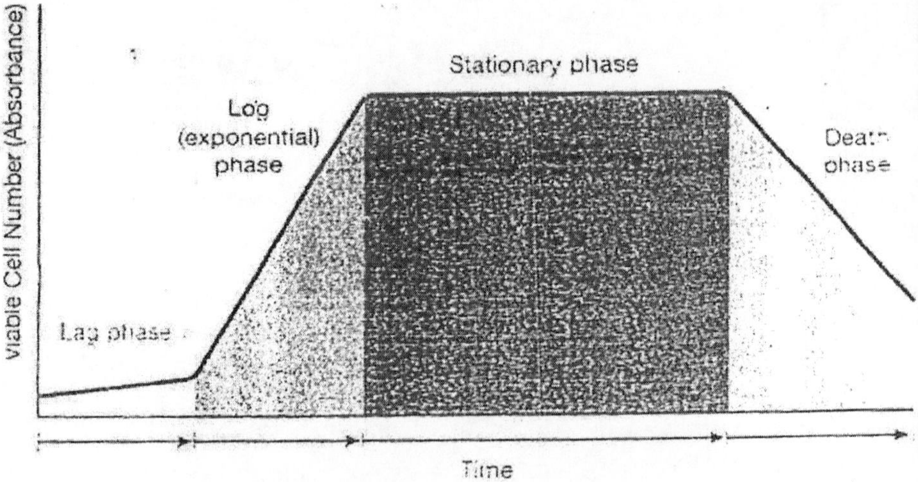

Escherichia coli Growth Curve

4 Isolation of Plasmid

Introduction

Bacterial plasmids are double-stranded closed circular DNA molecules that range in size from 1 kb to more than 200kb. They behave as accessory genetic units that replicate and are inherited independently of the bacterial chromosome. Plasmids are almost always purified from cultures (grown in liquid medium containing the appropriate antibiotics) that have been inoculated with a single bacterial colony picked from agar plate. Many of the currently used p\lsmid vectors replicate to such a high copy number that they can be purified in large yield from overnight grown cultures.

For isolation of plasmid. bacteria are recovered by centrifugation and lysed by any one of a large number of methods, including treatment with non-ionic or ionic detergents, organic solvents, alkali or heat. The methods of plasmid isolation explained the relatively small size and covalently closed circular nature of plasmid DNA. In commonly used methods to isolate smaller plasmids, the cells after addition of EDTA. in some cases, lysozyme, are exposed to detergent and lysed by boiling or treatment with alkali. These treatments disrupt base-pairing, cause the linear chromosomal DNA of host to denature. However the strands of closed circular plasmid DNA are unable to separate from one another because they are topologically intertwined. When conditions return to normal, the plasmid DNA rapidly renatures and completely native super helical molecules are reformed.

3.1. Protocol 1 (Standard Protocol)

Isolation of Plasmid by Alkali-lysis-Method

Requirements

(a) **Glassware and other materials:** Pasteur pipettes. spirit lamp, sterile microfuge tubes, 'rnicropipettes (P-20, P-200 and P- 1000), sterile micro pipette-tips, *etc.*

(b) **Instruments:** Microfu.qe, vortex mixer.

(c) ***E. coli:*** (Strain DH5a) having a plasmid with an antibiotic resistant gene, and/or *E. coli* (DH5a) with a recombinant plasmid

(d) **Reagents:**

1. Luria-Bertani (LB) medium
2. Ampicillin
3. STE buffer
4. Solution I (GTE buffer)
5. Solution II
6. Solution III
7. Ethanol (absolute and 70 per cent)
8. TE buffer

Procedure

A. *To Set Up an Overnight Culture*

1. Take 5 ml of LB medium to 25 or 50 ml conical flask or culture tube. Add 100 µg of ampicillM to the medium.
2. Inoculate it either with a liquid culture stored at 4°C (make 1 per cent inoculum) or with a glycerol culture. The best condition for inoculation is to inoculate with a single colony from a plate.
3. Let it grow overnight shaking at 37°C.

B. *Isolation of Plasmid*

1. Cool the overnight culture by keeping on ice.
2. Take 1.5 ml 0/N culture in a microfuge tube.
3. Centrifuge in a microfuge **at** full speed for 5 min. to pellet the bacteria.
4. Decant the supernatant carefully. Remove the remaining medium with the help of a micro pipette.
5. Resuspend the pellet in 100 III of ice-cold Solution I by vortexing vigorously.

   Note: Some strains of bacteria shed their cell wall components, which can inhibit the restriction enzyme digestion. To avoid this problem, the pellet is resuspended in 0.5 ml of STE buffer and centrifuge. After removal of STE buffer, resuspend in Solution I as above.
6. Add 200 µl of freshly prepared Solution II.
7. Close the tube and invert to mix the contents and keep at room temperature for exactly 5 minutes.
8. Add 150 µl of ice-cold Solution III and mix the contents and keep the tube on ice for 3-5 min.
9. Centrifuge at 10000 rpm for 15 mm at room temperature. Transfer the supernatant into a fresh tube.
10. Precipitate the DNA with two volumes of ethanol at room temperature. Allow the mix and keep at room temperature for 2 min.

11. Centrifuge at 10000 rpm for 10 min at 4 °C.
12. Remove the supernatant and place the tube in an inverted position on a paper towel to drain off excess ethanol.
13. Rinse the pellet in 1 ml of 70 per cent ethanol and centrifuge at 10000 rpm for 10 min.
14. Remove the supernatant and allow the pellet to air dry.
15. Dissolve the pellet in 50 pA of TE buffer and store at 4 °C till further use.
16. Analyse 5 µl on 1 per cent TAE/agarose gel

3.2. Protocol 2 (Alternative Protocol)

Isolation of Plasmid by Rapid Plasmid Isolation Method

Requirements

(a) **Glassware and other materials:** Conical flasks, culture tubes, spirit lamp. sterile microfug.e tubes, micropipettes (P-20, P-200), sterile rnicropipette tips, ice-box, sterile tooth picks,

(b) **Instruments:** Incubator shaker, microfuge, vortex mixer, water bath

(c) *E coli.* (strain DH5a) having a plasmid with an antibiotic resistance gene

(d) **Reagents:**

1. Luria-Bertani (LB) medium
2. Ampicillin
3. Lysis Buffer

Procedure

A. To Set Up an Overnight Culture

1. Take 5 ml of LB medium in a 25 or 50 ml conical flask. Add 106 lig of ampicillin per ml of LB medium.
2. Inoculate it either with a liquid culture stored at 4°C (make 1 per cent inoculum) or with a glycerol culture. The best condition for inoculation is to inoculate with a single colony from a plate.
3. Let it grow overnight on constant shaking at 37°C.

B. Isolation of Plasmid

1. Cool the overnight culture by keeping on ice.
2. Take 1.5 ml 0/N culture in a microfuge tube.
3. Centrifuge in a microfuge at full speed for 5 min at 5000 rpm to pellet thc bacteria.
4. In the meantime, prepare a boiling water-bath.

5. Decant the supernatant and remove the remaining medium with help of a micropipette.
6. Add 30 p.1 of lysis buffer and suspend the pellet either with mild vortexing or by pipetting with a yellow tip. Keep the lysate at room temperature for 5 minutes.
7. Keep the tube in boiling water-bath for exactly 1 min. and thereafter chill the tube on ice and keep it for atleast 10 min.
8. Microfuge the tube for 15 min at room temp to pellet the high mol. wt. nucleic acids and bacterial debris.
9. Take out the supernatant without disturbing the pellet into a fresh tube.
10. Analyse 5 µl in 1 per cent TAE/agarose gel.

Reagents

1. ***Lysis Buffer***

Tris-HC1 (1M - pH 8.0)	-	100 µl
EDTA (0.5 M - pH 8.0)	-	2 µl
Sucrose	-	15 mg
Lysozyrne	-	2 mg
Pancreatic RNase	-	0.2 mg
BSA	-	0.1 mg to 1 ml

Reagents

1. ***STE Buffer***

Sodium chloride	-	0.1 N
Tris-HCl (pH 8.0)	-	10mM
EDTA (pH 8.0)	-	1 mM

2. ***Solution I***

Glucose	-	50 mM
Tris-HCl (pH 8.0)	-	25 mM
EDTA (pH 8.0)	-	10 mM

Glucose is to be added separately autoclaved and to be added before use. Store the individual components at 4°C

3. ***Solution H***

Sodium hydroxide	-	0.2 N
SDS	-	1 per cent

4. *Solution III*

Glacial acetic acid - 29.5 ml

Add Potassium hydroxide pellets till the pH comes to 4.8. Adjust the volume to 100 ml with distilled water. Don't autoclave. Store at room temperature.

TE Buffer

Tris-HCl (pH 8.0) - 10 mM

EDTA (pH 8.0) - 1 mM

5 Isolation of Genomic DNA

5.1 Isolation of Genomic DNA from Mammalian Tissue

Requirements

(a) **Mammalian Tissue:** Tissue dissected out from mice/rat such as liver, brain. muscle *etc.* Store the tissue immediately at -20°C or in liquid nitrogen till further use.

(b) **Glass and plastic ware and others:** Small homogenizer, micropipettes (P-20. P-200, P-1000), pipette tips, microfuge tubes. oak ridge tubes, glass pipettes (10 ml), cheese cloth

(c) **Instruments:** Water bath at 37°C, cooling centrifuge, rotary mixer

(d) **Chemicals:**

1. Extraction Buffer
2. 20 per cent SDS
3. Proteinase K (20mg/m1)
4. Distilled and buffered saturated phenol
5. Chloroform - isoamyl alcohol (24:1)
6. 3M Sodium acetate (pH 5.2)
7. Ethanol (absolute and 70 per cent)
8. TE (pH 8.0)
9. RNase A (10 mg/m1)

Procedure

Day One

1. Homogenize the tissue in extraction buffer at 4°C and filter the homogenate through four layers of cheese cloth. Centrifuge at 4000 rpm for 10min at 4°C and remove the supernatant.
2. Suspend the pellet in 5-10 volumes of extraction buffer (about 5-10 for lg of tissue) at room temp.
3. Add Proteinase K to a final concentration of 100 ptg/m1 and SDS to a final concentration of 1 per cent.

4. Mix by gently inverting the tube and incubate overnight at 37°C.

   Note: The mixture can also be incubated at 50°C in a water bath atleast for 1 hour by periodic swirling.

Day Two

5. Add equal volume of buffered phenol, mix gently but thoroughly on a rotary shaker or by hand for 15 min and centrifuge at 10000 rpm for 10min.
6. Remove the viscous aqueous phase into a clean centrifuge tube using a wide-pore pipette. Take care not to disturb the white interface which is formed by protein precipitation. Repeat the step twice.
7. Mix the aqueous phase with equal volume of chloroform: isoamyl alcohol. Mix gently and centrifuge at 10000 rpm for 10 min and takeout the aqueous phase.
8. Add 1/30th volume of 3M sodium acetate and 2 volumes of chilled absolute ethanol. Keep at -20°C for 1 hr or in an ice bucket for 2 hours.
9. Carefully remove the precipitated DNA with a wide pore pipette tip into a microfuge tube containing chilled 70 per cent ethanol
10. Briefly spin, remove alcohol as much as possible, air dry and dissolve in appropriate volume of TE (200 µl -1000 µl). Allow 6-24 hr for the DNA to be completely dissolved.
11. To analyse the quality (fragment size) and quantity (concentration), run on 0.6 per cent TAE/agarose gel with lambda DNA as standard.
12. Add 2 µl of RNase. Incubate at 37°C for 2 hours.
13. Repeat the steps 5 – 11 to remove RNase.

Reagents

1. Extraction Buffer

Tris-HCl (pH 8.0)	-	100mM
EDTA (pH 8.0)	-	50 mM
NaC1	-	150mM
SDS	-	1 per cent

Make an SDS free stock solution and autoclave. Prior to DNA extraction, add 1/5 volume of 20 per cent SDS to an aliquot of this stock solution to make up the ready to use extraction buffer.

2. Proteinase K

20 mg/ml of Proteinase K is dissolved in autoclaved Millipore water and stored at 20°C in small aliquots.

3. *Buffered Phenol*

1. Melt the phenol crystals at 65°C. Add 0.1 per cent of 8-hydroxyquinoline (which is an antioxidant and gives a distinguishable yellow colour to phenol) to 500 ml of phenol (liquid) in a fume hood and mix on a magnetic stirrer.

   Note: If the phenol crystals appear reddish, distill the phenol before use.
2. Add 500 ml of 100 mM Tris-HC1 (pH 8.0), stir for 30 min at room temperature at low speed and stand to allow the complete phase separation.
3. Carefully decant out the aqueous phase.
4. Repeat the steps 2 and 3 until-the phenol phase is pH 8.0.
5. Add 250 ml of 50 mM Tris-HCl (pH 8.0) or TE buffer (pH 8.0):
6. Store at 4°C in brown glass bottles or clear glass bottles wrapped in aluminium foil. Phenol buffered in this way can be stored at 4°C for atleast 2 months.

4. *Chloroform : Isoamyl Alcohol (24:1)*

Mix 24 parts of chloroform to one part of isoarnyl alcohol. Chloroform alone can also be used for the extraction of DNA.

5. *3M Sodium Acetate (pH 5.2)*

Adjust pH to 5.2 with glacial acetic acid.

6. *TE Buffer (pH 8.0)*

Tris-HCl (pH 8.0) - 10 mM

EDTA (pH 8.0) - 1 mM

7. *RNase A (10 mg/m1)*

1. Dissolve DNase free RNase A in RNase buffer (10 mM Tris-HCl (pH 7.5), 15 mM NaC1).
2. Heat to 100°C for 15 min.
3. Allow to cool to room temperature slowly.
4. Store at -20°C in aliquots.

5.2. Isolation of Genomic DNA from Flies (Drosophila or Many other Insect Species)

Requirements

(a) *Drosophila* male flies

(b) **Glass and plastic ware and others:** Small homogenizer, micropipettes (P-20, P-200, P-1000), pipette tips, microfuge tubes, oak ridge tubes, glass pipettes (10 ml), cheese cloth

(c) Instruments: Water bath at 37°C, cooling centrifuge, rotary mixer

(d) Chemicals:

1. Extraction Buffer (100 mM Tris-HCl (pH 8.0), 50 mM EDTA (pH 8.0), 50 mM NaC1)
2. 20 per cent SDS
3. Proteinase K (20mg/m1)
4. Distilled and buffered saturated phenol
5. Chloroform - isoamyl alcohol (24:1)
6. 3M Sodium acetate (pH 5.2)
7. Ethanol (absolute and 70 per cent)
8. TE (pH 8.0)
9. RNase A (10 mg/ml)

Procedure

1. Homogenize adult male flies (– 50 in numbers) in 1.5m1 of extraction buffer at 4°C. (Females are generally not used because their abdomen contains a large proportion of nurse cells in which a part of the gnome is under replicated. Thus they do not represent true diploid genome).
2. Spin the contents at 4°C for 5min at 7000 rpm.
3. Resuspend pellet in 500 µl of extraction buffer with 1 per cent SDS. Add Proteinase K to a final concentration Of 100 µl/ml and incubate at 37°C for 2h with mixing of the contents of the tube by gentle tapping 2-3 times during this period.
4. Add equal volume of phenol and mix gently on a rotary shaker.
5. Centrifuge for 10 min at 10000 rpm at room temperature to separate the. viscous aqueous phase. Repeat the step twice.
6. Transfer the aqueous phase to a fresh tube and extract with chloroform-isoamyl alcohol. Centrifuge at room temperature for 10 min at 10000 rpm.
7. Collect the top aqueous phase into a fresh tube, add 1/10th volume of 3M sodium acetate and 2 V2 volumes of chilled ethanol to precipitate DNA. Leave at 200C for 1-2h or longer
8. Spin at 10000 rpm for 20 min at 4°C. Pour off the supernatant and wash the pellet with 70 per cent ethanol. Spin at 10000 rpm for 5 min at 4°C, pour off ethanol, air dry the pellet and dissolve in about 50 µl TE
9. Add RNase to a final concentration of 100 µg/ml. Incubate at 37°C for 2 h. Repeat steps 4 - 8 to remove RNase.

5.3. Isolation of Genomic DNA from Human Lymphocytes

Requirements

(a) Human whole blood

(b) **Glass and plastic ware and others:** Micropipettes (P-20, P-200, P-1000). pipette tips, microfuge tubes, *etc.*

(c) **Instruments:** Cooling centrifuge

(d) **Chemicals:**

1. Lysis Buffer [0.32 M sucrose, 1 per cent Triton X-100, 1 mM MgC12, 12 niM Tris-HC1 (pH 8.0)]
2. 20 per cent SDS
3. Proteinase K Buffer [0.375 M NaCl, 0.12 M EDTA (pH 8.0)]
4. Phenol-Chloroform (1:1)
5. 3M Sodium acetate (pH 5.2)
6. Ethanol (absolute and 70 per cent)
7. TE (pH 8.0)
8. RNase A (10 mg/m1)

Procedure

1. Take 1 nil of venous blood and add 1 ml lysis buffer. Mix yell and centrifuge at 10000 rpm for 10 min at 4°C.
2. Discard the supernatant and resuspend the pellet in 400 p.1 of lysis buffer and centrifuge at 10000 rpm for.5 min at 4°C and repeat the step once again.
3. Discard the supernatant and resuspend the pellet in 400 41 of distilled water. Centrifuge at 10000 rpm for 5 min at 4°C.
4. Discard the supernatant and add 100 tit of Proteinase K buffer and 10 IA 20 per cent SDS. Resuspend the pellet and mix it till frothing.
5. Add 140 pl NaCl and mix well. Add 400 p.1 distilled water and 400 III phenol-chloroform.
6. Mix well and centrifuge at 10000 rpm for 10 min at 4°C to ,separate the viscous aqueous phase.
7. Transfer the aqueous phase to a fresh tube and add I nil chilled ethanol to precipitate DNA. Leave at -20°C for I-2h or longer.
8. Spin at 10000 rpm for 20 min at 4°C. Pour off the supernatant and wash the pellet with 70 per cent ethanol. Spin at 10000 rpm for 5 min at 4°C, pour off ethanol, air thy the pellet and dissolve in about 50 ul TE.

5.4. Isolation of Genomic DNA from Plant Tissues

Requirements

A. **Plant tissue:** Broad range of tissues, fresh or dried including calli from different species

B. **Liquid nitrogen**

C. **Glass and plastic ware:** 50 ml oak ridge tubes, mortars and pestles (one per sample), funnels

D. **Instruments:** Water bath at 65°C, cooling centrifuge, rotary shaker

E. **Reagents:**

1. Extraction buffer (100 mM Tris-HCl (pH 8.0), 50 mM EDTA (pH 8.0), 500 mM NaC1, 10 mM p-me$_r$captoethanol)

   ***Note*:** Add P-mercaptoethanol before use.
2. 20 per cent SDS
3. 5M potassium acetate stock solution (Dissolve 122.75 gm of potassium acetate in 180 ml distilled water and make upto 250 ml with distilled water. Autoclave and store at room temperature).
4. 5M potassium acetate working solution (The working solution of potassium acetate is 5M with respect to acetate and 3M with respect to potassium. Mix 60 ml of potassium acetate stock solution and 11.5 ml of glacial acetic acid. Add 28.5 ml of distilled water and store at room temperature).
5. Isopropanol
6. 3M sodium acetate pH 5.2
7. 70 per cent ethanol
8. RNase solution (10 mg/ml)
9. Phenol-chloroform-isoamyl alcohol (25:24:1)
10. Chloroform-isoamyl alcohol (24:1)
11. TE buffer pH 8.0

Procedure

1. Weigh 0.5 to 1 g of tissue and freeze rapidly in liquid nitrogen and grind to fine powder with a mortar and pestle avoiding the tissue from thawing.

   ***Note*:** It is important not to let the tissue thaw once frozen, until it is added to the buffer.
2. Transfer the frozen powder into a 50 ml centrifuge tube containing 15 ml of Extraction Buffer and 10 mM13-mercaptoethanol.
3. Add lml of 20 per cent SDS and mix thoroughly by vigorous shaking and incubate the tubes at 65 °C for 10 min.

4. Add 0.5 ml of 5M potassium acetate working solution, shake vigorously and incubate on ice for 20 min.
5. Centrifuge at 10000 rpm for 20 min. Pour the supernatant through sterile small funnel containing two layers of gauze cloth into a clean 50 ml tube containing. 10 ml chilled isopropanol kept in ice bucket.
6. Mix gently by inverting the tubes several times and incubate at -20°C for 30 min. A DNA cloud should appear.
7. Centrifuge at 10000 rpm for 20 min. and gently pour off the supernatant and lightly dry the pellets by inverting the tubes on paper towels.
8. Dissolve the pellet in 0.7 ml of TE and kept at room temperature for 30 min.
9. Transfer the DNA to an Eppendorf tube and centrifuge at 10000 rpm for 15 min to remove insoluble debris.
10. Transfer the DNA to a new Eppendorf tube and add 75 ul 3M sodium acetate and 500 Al isopropanol. Mix well without vortexing and pellet the DNA by centrifuging for I - 5, min at 12000 rpm in a microfuge.
11. Discard the supematant and add 700 μl cold 70 per cent ethanol and centrifuge at 12000 rpm. Discard the ethanol and air dry the pellet and dissolve the pellet in 500 μl of TE buffer (pH 8.0) or deionised water.
12. 13:Add 10 ul of RNase solution, incubate for 2hrs at 37°C.
13. Add equal volume of phenol-chloroform-isoamyl alcohol and mix gently on a rotary shaker.
14. Centrifuge for 10 min at 10000 rpm at room temperature to separate the viscous aqueous phase.
15. Transfer the aqueous phase to a fresh tube and extract with chloroform-isoamyl alcohol. Centrifuge at room temperature for 10 min at 10000 rpm.
16. Collect the top aqueous phase into a fresh tube, add 1/10th volume of 3M sodium acetate and 2 -3. volumes of chilled ethanol to precipitate DNA. Keep at -20°C for 1- 2h or longer.
17. Centrifuge at 10000 rpm for 20 min at 4°C. Pour off the supernatant and wash the pellet with 70 per cent. ethanol. Spin at 10000 rpm for 5 min at 4°C, pour off ethanol. air dry the pellet and dissolve in about 50 ul TE.

6 Electrophoresis (Agarose Gel)

Introduction

The standard method used to separate and identify DNA fragments is electrophoresis through agarose gels. The location of DNA within the gel can be determined directly by visualizing ethidium bromide stained fluorescent bands in ultraviolet light (254 nm or 310 nm or 354 nm) using a transilluminator. The electrophoretic mobility of DNA through agarose gel is dependent on the following 4 parameters:

a. Molecular size of DNA
b. Agarose concentration
c. Conformation of the DNA
d. Applied current

Requirements

(a) Glassware and instruments: Conical flask, measuring cylinders, electrophoresis tank (horizontal), constant voltage power-supply, UV -transilluminator, micropipette (P-20), sterile micro pipette-tips, hotplate/water bath/microwave oven. *etc.*

(b) Chemicals:

1. TAE buffer
2. Ethidium bromide
3. Agarose
4. Loading dye

Procedure

A. *Preparation of Submarine Agarose Gel (1 per cent)*

1. Prepare a gel mould by wrapping scotch tape around a gel tray or a glass plate of desired dimensions.
2. Place the gel tray on a flat surface and assemble the comb (slot former). The comb is fixed in such a way that there should be a gap of 1-2 mm between the surface of glass plate and the tip of the teeth in the comb.

3. Calculate the volume of gel solution to be poured from the dimensions of the gel tray or glass plate and the desired thickness of the gel.
4. Weigh the agarose (to make 1 per cent gel) and transfer to a conical flask with 1X TAE buffer.
5. Keep the agarose in boiling water-bath/hot plate/microwave oven for 15-20 min. to melt the agarose.
6. Cool down the gel solution approximately to 45-50°C; and add ethidium bromide to a final concentration of 0. 5 µg/ml
7. Mix the gel solution avoiding air-bubbles, and pour in the gel mould.
8. Let it solidify. The time of solidification may vary from 15-30 min 'depending upon the room-temperature. Once solidified, it acquires a: faint bluish-white tinge.
9. Remove the comb carefully, and then remove the tape also carefully. (The agarose gel does not stick to the tray/glass, and is also very brittle; hence do not let the gel slide from gel tray/glass-plate.

   Note: Use gloves while handling ethidium bromide and solutions containing EtBr.
10. Transfer the gel to a horizontal electrophoresis tank with 1X TAE. Inspect the wells. If it contains any small pieces of agarose, flush it out with the help of a syringe filled with tank-buffer.

B. Loading of DNA and Electrophoresis

1. Mix the components in a microfuge tube in the following order.

   Plasmid - 5 µl
   H_2O - 15 µl
   Loading dye - 5 µl
   Total - 25 µl
2. Connect the electrodes; cathode - at the origin and anode - to the other end of the gel. (DNA being negatively charged runs from cathode to anode).
3. Perform electrophoresis at 100 volts till the bromophenol blue migrated 2 cm from the bottom of the gel.
4. Visualise the DNA band in a UV-transilluminator at 300 nm. Note: Use face-mask while visualising the gel in transilluminator.
5. Photograph the gel for record in a gel-doe or using a red filter. Also, make a tracing of the band on a transparent cellophane sheet for immediate discussion.

Reagents

1. 50X TAE buffer

Tris base - 242

Glacial acetic acid - 57.1 ml

EDTA (pH 8.0) - 100 ml

Distilled water to make up the volume to 1000 ml.

2. *Ethidium Bromide*

10 mg/ml in water (stored in dark at 4°C)

3. *Loading Buffer (Tracking dye)*

Bromophenol blue - 0.25 per cent

Xylene cyanol - 0.25 per cent

Sucrose - 40 per cent

7 Quantification of DNA

Introduction

Quantification of DNA is a very important step in many procedures where it is necessary to know the amount of DNA that is present when carrying out restriction digests or performing different techniques such a.A PCR and RAPDs. There are several methods for quantifying DNA, the well known being

1. The comparison of an aliquot of the extracted sample with standard DNAs of known concentration using gel electrophoresis and
2. Spectrophotometric determination.

With both the methods the quality and purity of the extracted DNA is obtained. Normally both methods are used, but if the amount of DNA available is small the gel electrophoresis technique alone may be employed. If there is no limitation of DNA, spectrophotometric measurement should also be taken. Following this, the samples may be adjusted to the same concentration, and this can then be confirmed by gel electrophoresis.

Requirements

(a) Instruments: Electrophoresis equipment, Microwave oven or magnetic stirrer hot plate or boiling water bath. UV transilluminator, photographic equipment (Polaroid system or gel documentation system), spectrophotometer and UV transparent quartz cuvettes, *etc.*

(b) Agarose (standard melting temperature), 50x TAE stock solution (working solution is 1X), 6X loading dye, standard DNA (*e.g.* X phage DNA digested with *HindlII* enzyine, *etc.*

Procedure

1. Electrophoresis of a DNA Sample of Unknown Concentration with a known Standard

1. Prepare a 0.8 per cent TAE/agarose gel as mentioned in protocol V, steps, 1- 10.
2. Mix 3 1.1.1 distilled water, 1 ul loading dye and 2 1 DNA and load on to the gel.

3. Load 2 jig of DNA digested with *Hitic/III.*
4. Run the gel at 60-70 V until the dye is 3-4 cm from the well.
5. Illuminate the gel with UV light and photograph the gel under UV.
6. Compare the intensity of DNA bands of the samples with the intensity of X bands. As the amount of DNA in each band of X is known, the amount of DNA of each sample can be calculated by comparing the fluorescent yield of the sample with those of the X bands.

λ DNA is approximately 50 kb and when digested with HindIII, eight bands of different. sizes appear in the gel. Taking into account that 2 jig of DNA were loaded - onto the gel, the amount of DNA in each band will be as follows

Size of the Band of λ (in bp)	*Approx. Concentration (in ng)*
23130	900
9416	370
6557	260
4361	180
23??	90
2027	80
564	?0
125	5

2. *Spectrophotometric Determination of DNA Concentration*

Dilute 1.5 j.il of DNA to 1500 111 with distilled deionised water and read the OD at A230, A260 and A280. The A260/A280 ratio provides an estimate of the purity of the nucleic acid. In a pure DNA sample this ratio should be around 1.8. Lower values indicate protein or phenol contamination. A230 should be less than A260 and may be the same as A280. High A230 readings indicate that residual phenol remains in the preparation.

An A260 of 1 corresponds approximately 50 ig/ml of double stranded DNA in a 1 cm quartz cuvette. The nucleic acid concentration is calculated as follows:

A260 X 50 mg/ml X 0.001 1.11/m1 X dilution factor (1500 01.5 1i.1) = X iig/p.1

8 Polymerase Chain Reaction

Introduction

The polymerase chain reaction (PCR), which was invented by Kary Mullis, is a primer-mediated enzymatic amplification of specifically cloned or genomic DNA sequences. A therrnostable DNA polytnerase, *Tag* DNA polymerase, catalyzes the buffered reaction. The reaction contains an excess of an oligonueleotide primer pair and four deoxynucleoside triphosphates (dNTPs), used to make millions of copies of the tens or tens of thousands of nucleotides long target sequence contained in the template DNA.

In molecular biology and genetic disease research for the identification of new genes, PCR is widely used. Viral targets, such as HIV-1 and HCV can be identified and quantified by PCR. Active gene products can be accurately quantitated using RNA-PCR. Owing to PCR, there have been revolutionary advancements in the fields of anthropology and evolution as sequences of degraded ancient DNAs can be tracked very easily. With its exquisite sensitivity and high selectivity, PCR has been used for wartime human identification and validated in crime labs for mixed-sample forensic casework. In the realm of plant and animal breeding, PCR techniques are used to screen for traits and to evaluate living four-cell embryos. High sensitivity quantisation and identification of environmental and food pathogens can also be carried out.

PCR Process

The PCR process requires a repetitive series of the three fundamental steps that defines one PCR cycle: double-stranded DNA template denaturation, annealing of two oligonucleotide primers to the single-stranded template. and enzymatic extension of the primers to produce Copies that can serve as templates in subsequent cycles. The target copies are double-stranded and bounded by annealing sites of the incorporated primers. The 3' prime end of the primer should complement the target exactly, but the 5' end can actually be non-complementary tail with restriction enzyme and promoter sites that will also be incorporated. As the cycle proceed, both the original template and the amplified - targets serve as substrates for the denaturation, primer annealing, and primer extension process. Since every cycle theoretically doubles the amount of target copies, a geometric amplification occurs. Given an efficiency factor for each cycle, the amount of amplified target, Y. produced from an input copy number, *X*, after *n* cycle S is

$$Y = X\ (1+ \text{efficiency})^n$$

With this amplification power, 25 cycles could produce 33 million copies. Every extra 10 cycles produces 1024 more copies. Unfortunately,_ the process becomes self-limiting and amplification factors are generally between $10^{)}$ – and 10^{9} – fold. Excess primers and dNTPs help drive the reaction that commonly occurs in 10mM Tris-HCl buffer (pH 8.3 at room temperature). In addition. 50 mM Cl is present to provide ionic strength and magnesium on is required as an enzyme cofactor.

The denaturation step occurs rapidly at 94-96°C and the annealing of the primer depends on the T_{rn}. or melting temperature, of the primer : template hybrids. The best annealing temperature is determined by optimization. Extension occurs at 72°C for most templates.

PCR: (Flowchart)

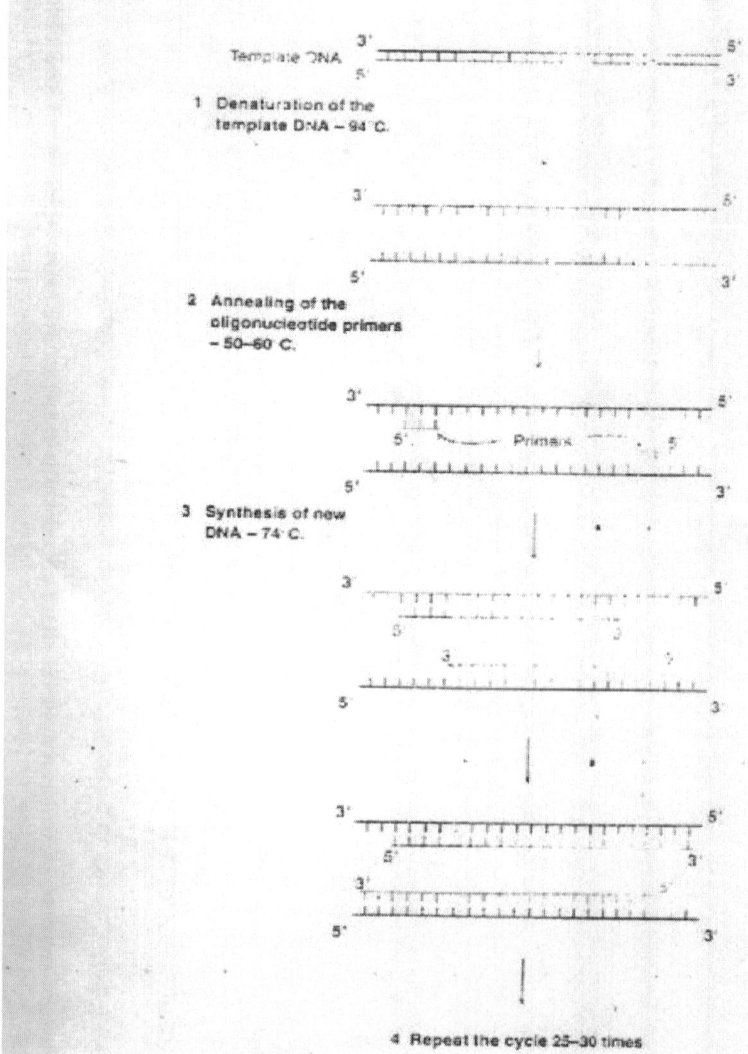

Considerations for Amplifying DNA using PCR

1. The most critical step in PCR is primer selection. The best way to select a primer is to consider the following points :
   A. The last three 3' nucleotides of both primers should not be able to hybridize to them-selves or to the other primer. If this occurs, the primers will form primer dimers in solution, the polymerase will extend these dimers and essentially create new primers which will be non-specific.
   B. Match the melting temperatures of the two primers. A simple rule for determining the T, 6f a primer is the (2(AT) + 4(GC)) rule or 2°C for every A Or T nucleotide and 4°C for every G or C nucleotide. A good range of T,,, to stay within is 50-60°C.
   C. The typical parameters for a cycle in PCR are the following:

      10 sec – 1 min. at 94°C

      30 sec – 2 min. at the primer annealing temp and 1 –3 min at 72°C

      Note: The length of time varies depending on the type of tube/titre plate used to do the PCR.
   D. The number of cycles to be used **in** a single reaction varies according to the type of amplification. Typical PCRs use between 20-40cycles.

      Remember that, as one increases the number of cycles, possibility of obtaining a mutation in the final PCR product increases.

      Always: Remember to sequence your final PCR product (*i.e.* the clone).
   E. A good initial step in amplifying a fragment of DNA is to vary the buffer conditions and hold all other components constant.

Initial concentrations for the other components are the following:

200mM dNTPs,.1mM oligo A, 1mM oligo **B.** lricz template DNA/10m1 of reaction volume (assuming template is 5 kb in length)

A good buffer series to try is the following **(all** given concentrations are 10X):

1. 100 mM Iris pH 8.4, 500 mM KC1 and 15 mM $MgCl_2$
2. 100 mM Tris pH 8.8, 470 mM KC1 and 27mM $MgCl_2$
3. 100 mM Tris pH 8.4, 500 mm KC1 and 20 mM $MgCl_2$
4. 100 mM Tris pH 8:8, 500 mM KC1, 60 mM $MgCl_2$ and 10 mM DTT
5. 100 mM Tris pH 8.4, 500 mM KCI and 60 miM $MgCl_2$
6. 100 mlq Tris p1-I 9.0, 400 mM $(NH_4)2SO_4$ and 30mM MQC_{12}

Once one has established which buffer is best, the concentration of the other components can be varied in order to maximize the efficiency of the reaction. This step is usually not necessary for most PCRs, however, for sensitive amplifications it may be necessary.

For the efficient amplification of target sequences individual reaction components, time and temperature parameters must be adjusted. The following suggestions may help to carry out gene amplifications successfully.

1. *Sample Volume and Microfuge Tube*

Most amplifications are performed at 50 and 100 ill volume in 0.5 m1/0.2 ml microfuge tubes. Larger volumes are not recommended as there will not be adequate thermal equilibrium of the reaction mixture.

2. *Template DNA*

Typically nanogram amounts of cloned template, up to microgram amount of genomic DNA or 20.000 target copies are chosen to start optimization trials. Higher the amounts the template DNA inhibits or results in non-specific amplifications.

3. *Primers*

Primers are oligonucleotides, typically 15 to 30hases long. It is important that the primers do not contain more than two bases complimentary with each other, especially at their 3' ends to avoid the formation of Primer Dimer. There should not be any internal secondary structure. Typically, 40 to 55 per cent G+C content is often recommended. Optimal annealing temperature must be determined empirically. The concentration of the primers is typically within 0.1 to 1 iM range. The melting temperature (TM) of forward and reverse primers should be the same.

4. *Deoxynucleotidetriphosphates*

The final concentration of each dNTP in a standard amplification reaction mixture is 200 ti.M. It is important to keep the four dNTP concentrations above the estimated Km of each dNTP (10 to 15 n.M) • and balanced for best base incorporation fidelity.

5. *10X Buffer*

The 10X buffer contains 600mM Tris-HCl (pH 8.5), 15mM $MgCl_2$, 250 mM KCl, 100 mM13-mercaptoethanol and 0.1 per cent Triton X-100.

6. *Taq DNA Polymerase*

Taq DNA polymerase is a 94 KD thermostable DNA polymerase which lacks 3 'to 5' exonuclease activity but has 5' to 3' exonuclease activity in addition to 5' to 3' polymerase activity. In a standard aliquot of Taq DNA polymerase for a 100 1.11 reaction, there are about 1010 molecules. For most amplification reactions 1.5 to 2.5 units of enzyme are recommended as excess of it leads to non-specific amplification.

7. *Thermal Profiles*

DNA denaturation is the critical step in the DNA amplification reactions. The time specified for DNA depends on various factors like, nature of template,

GC content, secondary structure *etc.* For most amplification incubation time for DNA denaturation is I minute at 92 to 95°C, with 94°C being standard choice, Primer annealing in most of the cases is done between 45° to 60°C. The specified annealing temperature is calculated empirically and is usually 5°C less than the melting temperature (Tm) of the primers used. Extreme annealing temperature may result in non-specific amplification. Primer extension. in most amplifications occurs efficiently at a temperature of 72°C and seldom needs optimization. Time specified for extension depends on the length of target sequence. For example for 1 kb target DNA usually 1 minute extension is recommended.

Material

1. Tag DNA polymerase
2. Deoxynucleotidetriphosphates (supplied as 10mM solution)
3. 10X Taq polymerase buffer
4. 10X Taq polymerase Buffer without Magnesium Chloride. It contains all above components except MgC12, which is supplied separately
5. DNA template
6. Primers
7. Deionised distilled water

Procedure for Control DNA Amplification

Add the reagents in the following order.

1. Add 39 1.11 of sterile MilliQ water Or autoclaved double distilled water into a 0.5 ml PCR tube.
2. Add 5 IA of 10X Taq polymerase assay buffer with magnesium chloride.
3. Add 2 lii of 2.5 mM dNTP mix solution.
4. Add 2 i.t1 of template DNA (100 ng 4.11).
5. Add 1 il each of forward and reverse primers (250 ng/.i1)
6. Add 0.5 to 1 of Taq DNA Polymerase (5 Units/p.1)
7. Mix the solution gently and spin briefly to settle down the contents.
8. Layer the reaction mix with 50 of mineral oil to avoid any evaporation.
9. Carry out the amplification using following reaction condition for at least 30 cycles.

   Denaturation at 94° C for 30 seconds

   Annealing at 48°C for 60 seconds

   Extension at 72° C for 2 minutes
10. After the reaction is over take out the reaction mix and run 5 p.1 of the aqueous layer in 1 per cent agarose gel. Visualize under UV light for 0.8 kb DNA fragment.

Note: The above protocol is recommended for any routine Amplification reaction.

Tips for Successful DNA Amplification

The ideal way to perform amplification reaction is in a dedicated room, using reagents and equipment also dedicated only to amplification reaction. An attempt should be made to keep concentration stocks of target sequences away from amplification area and equipment. Because of the power of the procedure, meticulous care must be taken to avoid carry-over of DNA from one tube to another to prevent false positives.

1. If there is no yield of the desired product,
   a. One should determine that the enzyme is being added reproducibly preferably in a master mix.
   b. Also one should ensure complete denaturation with each cycle by using properly sized 0.5 ml microfuae tubes and by allowing sufficient time at the denaturation plateau temperature.
   c. Consider a slightly higher denaturation temperature.
   d. Check the chemical integrity of the primers.
   e. Examine need to pre-incubate at 95°C for 5 to 10 minutes in the absence of enzyme to inactivate harmful proteases or nucleases in the sample.
2. If a smear of amplified products is seen on the agarose gel, reduce *Taq* DNA polymerase concentration; increase annealing temperature; reduce the magnesium ion concentration; minimize the incubation times of the annealing and extension steps; decrease the number of cycles; and run molecular weight marker to check gel electrophoresis.
3. If a primer-dimer artifact band is seen on the agarose gel, check primer sequences for 3' complementarily; design longer primers; increase the amount of target DNA; reduce primer concentration; reduce number of cycles; and raise the annealing temperature.

9 Cloning of PCR Product using T/A Overhang

The cloning of PCR product into an appropriate vector allows identification of the product' by sequencing and characterisation of the product's transcript by northern hybridisation analysis or ribonuclease protection assay. PCR products may be cloned into the vector by cohesive or blunt end ligation. Cohesive end ligation requires the addition of restriction sites to the PCR primers. This requires additional enzymatic manipulation and purification steps before the product is cloned into a vector. Similarly. PCR products amplified with *Tag* polymerase must undergo further enzymatic manipulation before blunt end ligation. In addition, blunt end ligation is a less efficient process than cohesive end ligation.

The product from completed PCR reaction can be transferred directly to the ligation reaction mixture. This procedure is based on the fact that *Tag* polymerase has a nontemplate-dependent activity that adds a single deoxyadenosine- to the 3' ends of the double stranded DNA. The pTZ57R/T vector (MBI Fermentas) or PCR^{Tm} vector (Invitrogen) is a linearised molecule with 3' T-overhangs on either side of the insert site. This allows the cohesive end ligation of the PCR product.

The steps followed in this protocol are essentially from MBI Fermentas InsT/A PCR product Cloning Kit in order to clone a 1.8 kb fragment of HSP60 into pTZ57R/T vector. The high efficiency of the kit is based on the use of a specially designed cloning vector, pTZ57R1T. The vector has been pre-cleaved with *EcoRV* and treated with terminal dexynucleotidyl transferase to create Y-ddT overhangs at both ends. When a PCR fragment with 3'dA overhangs is ligated into the vector, a circular molecule with two nicks is produced. The circular product can be used to directly transform *E. coli* cells with high efficiency.

The following table shows the structure of 3' ends of PCR products generated by *Tag* polymerase

5' end nucleotide of primer 3"-end nucleotide of product on complementary strand

A	T	- T,+A
C	G	+G>+A>>+C
G	C	+A>+C
T	A	+A (at a very low efficiency)

Procedure

1. *Ligation*

The efficiency of ligation is known to dependent on the purity of the PCR fragments. If the PCR product is sufficiently clean (a_homogenous band of desired size is observed on the gel), it can be directly used in the ligation reaction.

i. Dissolve the purified PCR fragment in 10-20 pi of TE buffer, determine the approximate DNA concentration by agarose gel electrophoresis and visual comparison to a known amount of DNA markers.

ii. Calculate the amount of PCR fragment required for ligation (in equimolar concentration) using the formula –

X ng of PCR product to ligate = (Y· bp of PCR product)(50 ng of vector)/(size in bp of the vector)

Three times "*X*" ng would be used for a 1:3 molar ratio.

iii. Add the following components into a 1.5 ml microfilize tube:

Plasmid vector pTZ57RYT vector (0.165 pg)	1 µl
PCR fragment (approx 0.495 pg)	2 µl
10 X ligation buffer	1 µl
PEG 4000 solution	1 µl
Deionised water	To 10 µl
T4 DNA ligase, 5li	1 µl

iv. Incubate at 22°C for I hr. For maximum yield, the reaction time can be extended to overnight.

v. A control ligation reaction can be performed using a control PCR fragment provided with the kit.

2. *Transformation*

Competent cells of the *E. coli* strains such as JM107, XL1-Blue, DH5a., SURE, NM522 can be used for transformation.

The following mutations are important for transformation:

en⁻clAl (mutation in the DNA specific endonuclease 1 gene), it helps to increase the yield and quality of plasmids;

recAl (mutation in general recombination gene);

lacZAM15 (partial deletion of 13-galactosidase gene), this mutation is required for blue/white selection;

lad.I (mutation leads to high levels of the *lac* repressor protein, inhibiting background transcription from the lac promoter).

3. Transformation and Selection of Positive Clones

Follow protocol 9.5

10 Cloning and Expression of Cloned Gene

Strategies for Cloning in Plasmid Vectors

In principle, cloning in plasmid vectors is very straight forward. The plasmid DNA is cleaved with a restriction enzyme and joined *in vitro* to foreign DNA, the resulting recombinant plasmids are then used to transform bacteria. In practice, however the plasmid vectors must be carefully chosen and processed to minimise the effort required to identify and characterise recombinants. The major difficulty is distinguishing, between plasmid that contains inserted foreign DNA vector molecules that have recircularized without insertion of foreign DNA.

The, easiest fragment to clone carries non-complementary protruding termini generated by digestion with two different restriction enzymes. Since most of the present day vectors contain polylinker that has multiple cloning sites, it is almost always possible to find restriction sites that are compatible with the termini of the fragment of foreign DNA. The fragment of foreign DNA is then inserted in to the veqtor by a process known as directional cloning.

Fragment of foreign DNA carrying identical termini (either blunt-ended or protruding) must be cloned in a linearized plasmid vector bearing compatible ends. During the ligation reaction, the foreign DNA and the plasmid have the capacity to circularise and to form tandem oligomers. It is therefore necessary to carefully adjust the concentrations of the two types of DNA in the ligation reaction to optimise the number of – 'correct' ligation products. In addition, removal of the 5'-phosphte groups with alkaline phosphatase is frequently used to suppress self ligation and circularization of the plasmid DNA.

10.1 Restriction Digestion of Plasmids and Calculation of the Base-Pair Length of Plasmid and Insert using Lambda Phage DNA Digested with Hind III Restriction Enzymes as a Marker

Requirements

(a) **Glassware, Instruments, and other materials**: Electrophoresis tank (horizontal/submarine), constant voltage power-supply, UV-transilluminator. Sterile microfuge tubes, micropipettes (P-20. P-200 and P-1000), sterile micro pipette-tips, ice-box, microfuge, vortex mixer, water bath. *etc.*

(b) Chemicals:

1. Agarose, 50X TAE stock buffer, Ethidiurn bromide (10 mg/m1), loading dye
2. 10X 'Restriction enzyme buffers – as supplied by the manufacturer (Stored frozen at -20°C)
3. Restriction endonucleases like *EcoRl. Hindu, etc.* (st
4. DNA digested with *Hinc1111*
5. Isolated plasmids (with and without inserts or foreign I.

Procedure

A. Digestion of Plasmid to Linearise the Vector and Release the Insert

1. Mix following components in the order mentioned below in a microfuge tube on ice:

   i. To linearise the plasmid

   Plasmid DNA (1-2 pg) - 5 μl
   10X RE buffer - 2 μl
   Restriction enzyme 1 (*e.g. EcoRI)* - 2 μl
   Deionised distilled water - 20 μl

   ii. To release the insert:

   Plasmid DNA (1-2 pg) - 5 μl
   10X RE buffer - 2 μl
   Restriction enzyme 1 (*e.g.* EcoRI) - 2 μl
   Restriction enzyme 2 (*e.g.* Hind III) - 2 μl
   Deionised distilled water - 20 μl

2. Mix the contents by gentle tapping and puke spin in a microfuge to bring down all the liquid to the bottom of the tube and incubate at 37°C for 1 h.
3. Heat at 6'5°C for 10 min to stop the reaction, chill on ice. Add 5 ul of loading dye, mix and pulse spilt
4. Load the digested products on 1 per cent TAE/agarose gel as in Protocol V with λ DNA digested with *Hindill* as marker.
5. Visualise the DNA band in a UV-transilluminator. Photograph the gel for record using a red filter. Also, make a tracing of the band on a transparent cellophane sheet for immediate calculation and discussion.

B. *Calculation of Base Pair Length of Plasmid*

1. From either photograph or the tracing measure the distance of migration of fragments obtained from X. DNA digested with Hind III.
2. Plot a standard curve on semi-log paper by taking bp length of DNA fragment on Y axis and distance migrated on X-axis.
3. Also measure the distance of fragments obtained from plasmid digest and extrapolate the bp length *of* these fragments from the standard curve.

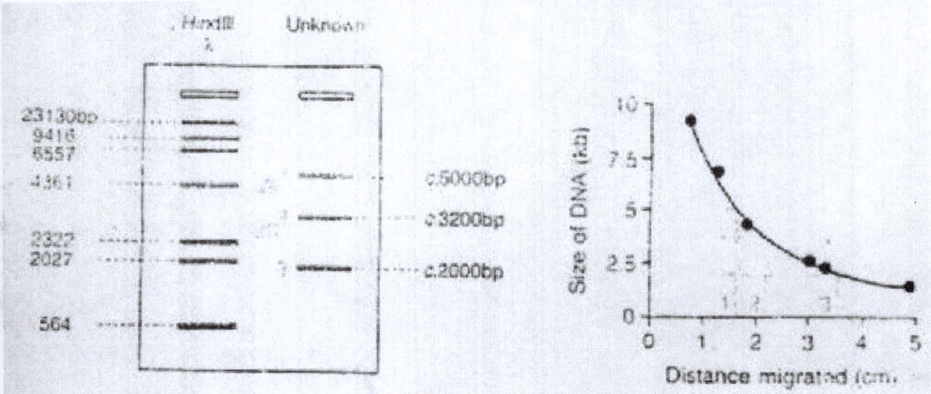

Figure showing the calculation of length of digested plasmid products

10.2. Gel Elution of the Restriction Digested Products

(As per manufacturer's instructions using the EZNA Gel Extraction Kit)

See Appendix.

10.3. Preparation of Vector DNA and Foreign DNA Fragment for Cloning

Requirements

1. Same as in 10.1
2. Calf Alkaline Phosphatase (CIP), and 10X CIP buffer
3. Buffer saturated phenol, phenol-chloroform, chloroform
4. 3 M sodium acetate (pH 7.0), 3 M sodium acetate (pH 5.2), 10 M Ammonium acetate, 0.5 M EDTA (pH 8.0)

Procedure

A. *Preparation of Phosphatase-Treated Vector*

1. Digest the plasmid prepared by miniprep with the desired restriction endonuclease (as in protocol 10.1)

2. Check 5 ul of the above plasmid for completion of digestion in 0.8 per cent agarose gel. Use undigested plasmid and marker DNA for comparison.
3. When the digestion is complete, proceed for phosphatase treatment as follows:

Digested plasmid	15 µl
10X CIP buffer	5 µl
H_2O to	49 µl
Calf Alkaline phosphatase (1 U)	1 µl

4. Incubate at 37°C for 30 min. Add another 1 [II of CIP and continue the incubation for 30 min.
5. After 1 h of CIP treatment, add 1 p.1 of 0.5 M EDTA (pH 8.0) to get a final conc. of 5 nil\/1. Incubate at 75°C for 10 min. to inactivate the CIP.
6. Cool the reaction to room temperature and extract once with equal volume of phenol. and once with equal volume of phenol-chloroform mixture.
7. To the aqueous phase add 0.1 volume of 3 M sodium acetate (pH 7.0), and 2 volumes of ethanol, mix well and precipitate the linear dephosphorylated vector at -20°C for 30 min.
8. Recover the DNA. by centrifugation at 4°C in a microfuge, wash the pellet with 70 per cent ethanol, air dry and dissolve in 10 µl of distilled water.
9. Run an aliquot of both the vector and the insert DNA to estimate the concentration of DNA in gel before setting up the ligation.

10.4. Ligation of Vector and Foreign DNA Fragment

Ligation of a segment of foreign DNA to a linearized plasmid vector involves the formation of new bonds between phosphate residues located at the 3'-hydroxyl moieties. Ligation of one end of DNA to another can be regarded as a bimolecular reaction whose velocity under standard conditions is determined solely by the concentration of compatible DNA termini. This is true no matter whether the termini are located on the same DNA molecule (intramolecular ligation) or on different molecules (intermolecular ligation). At low concentrations of DNA, recircularization of the plasmid DNA will occur with high efficiency. If the concentration of DNA in the ligation reaction is increased, a given end of the DNA is more likely to encounter a terminus located on another molecule before intramolecular ligation occurs. At high DNA concentrations therefore, the initial product of ligation will be dimmers and larger oligomers of the plasmid.

Requirements

1. Purified Vector and foreign DNA fragment
2. T4 DNA ligase and 10X ligase buffer

Procedure

1. Set up ligation reaction in total volume of 10 pi as follows (Ligation reaction (for a foreign DNA fragment that has length equal to vector DNA).

Ingredient	***Volume (μl)***	***Final conc.***
Vector DNA (100 ng4tl) with dephosphorylated5' termini	2	200 ng.
Foreign DNA fragment (100 ng./p,1) with compatible phosphorylated termini	2	200 ng (equimolar to vector)
10x ligase buffer with 10 mM ATP	1	1X
T4 DNA ligase	1	0.1 Weiss Unit
Deionised distilled water	to 10	

2. If the foreign DNA is smaller than vector, reduce the concentration of foreign DNA accordingly to bring it equal to molar concentration of the vector.
3. Include necessary controls like; ligation without foreign DNA fragment (Vector recircularization control), ligation of linear vector only that was not treated with phosphatase (Ligation control), *etc.* In each case, adjust the volume to 10 μl with H_2O.
4. Incubate the reaction for 4-16 h at 16°C.
5. Use 2 pl of the ligation reaction for transformation of bacteria. Store the remaining ligation reaction at -20°C for future use.

10.5. Transformation

When the bacteria are treated with ice-cold solutions of $CaCl_2$ and then briefly heated, they could be transfeeted with plasmid DNA. Apparently the treatment induces a transient state of "competence" in the recipient bacteria, during which they are able to take up DNAs derived from a variety of sources. Most commonly used methods yield transformants at a frequency of 10^7-10^8 transformants/p.g of supercoiled plasmid.

Requirements

(a) Glassware, Instruments, and other materials: Incubator, Gyro-shaker, sterile 50 ml and 15 ml polypropylene centrifuge tubes, and rest as mentioned in earlier experiments.

(b) Chemicals:

1. LB broth

2. **LB** agar plates with and without ampicillin – For plates with ampicillin, add ampicillin to final concentration of 100 µg/ml before plating.
3. **Calcium chloride** - 0.1 M (prepare 1 M stock solution of $CaCl_2$ and store 10 ml aliquots frozen at -20°C. Just before use dilute an aliquot to 100 ml with sterile water and by filter through 0.45 micron filter, and then chill on ice for use.

Procedure

Day I: Selection of E. coli DI-15a. on LB agar plates

1. Streak either from glycerol culture or from any viable culture stored at 4°C with platinum loop on LB agar plates to isolate a single colony.
2. Let the colony grow 0/N in an incubator at 37°C.

Day 2: Preparation for 0/N Culture

1. Inoculate 5 ml of LB medium with a single colony from LB agar plate.
2. Let it grow 0/N shaking at 370C.
3. Keep the following in the cold room for next day use - 50 ml centrifuge tubes, microfuge tubes and pipettes. 10 and 5 ml glass pipettes. $CaCl_2$ solution.

Day 3: Preparation of Competent Cells

1. Inoculate 100 ml LB medium in 1 litre flasks with 0/N culture (make 1 per cent inoculum).
2. Let the flask grow at 37°C shaking for 2-3 hrs. Measure OD at A550 after every 30 min. to find out when A5i0 in between 0.4 – 0.5 (*i.e.* when the cells are in early log phase).
3. Chili the flasks in ice for 5-10 min. Transfer the culture to pre-chilled 50 ml centrifuge tubes. Centrifuge at 5000 rpm in 5 min. at 4°C.
4. Decant the supernatant and suspend the pellet gently with the help of a pre-chilled glass pipette in 20 ml of 200 mM $CaCl_2$ and incubate in ice-water for 20 min.
5. Centrifuge at 5000 rpm in 5 min. at 4°C.
6. Decant the supernatant and suspend the pellet gently in 4 ml of 80 mM $CaCl_2$ solution.
7. Aliquot the cells in microfuge tubes either in 200 ul or its multiples, and immediately freeze in liquid nitrogen and store the cells at -70°C.
8. If not used immediately, the competent cells may be stored on ice 0/N.

Day 4: Transformation

1. Thaw a 200 µl aliquot of competent cells on ice.
2. Add 2 µl of ligation mix containing approximately 50-100 ng of DNA to the competent cells. The volume of DNA/ligation mix to be added to a 200 ill aliquot of competent cells should not exceed 10 µl. Mix the contents of tube by swirling gently. Incubate on ice for 30 min.
3. Heat shock at 42°C for exactly 120 seconds. Do not shake the tubes.
4. Rapidly transfer the tubes to an ice bath. Allow the cells to chill for 1-2 minutes.
5. Add four volumes (0.8 ml) of LB medium and keep in a water bath at 37°C for 1 hr to allow the bacteria to recover and to express the antibiotic resistance marker encoded by the plasmid.
6. Centrifuge at 5000 rpm for 5 min • to settle down the cells. Remove medium to maximum. Resuspend the cells in the remaining medium and plate on the LB ampicillin plates with IPTG and X-gal.
7. Include following controls
   a. No DNA control
   b. Transformation efficiency control.
8. Incubate the plates at 37°C O/N.

10.6. Identification of Bacterial Colonies that Contain Recombinant Plasmids by cc-complementation

There are four methods that are commonly used to identify bacterial colonies that contain recombinant plasmids

1. Restriction analysis of small scale preparation of plasmid DNA
2. Insertional inactivation
3. Screening by hybridisation
4. cc-Complementation

1. Restriction Analysis

In restriction analysis, a number of independently transformed bacterial colonies are picked and grown in small cultures. Plasmid DNAs isolated from each culture are analysed by digestion with restriction enzyme and gel electrophoresis.

2. Insertional Inactivation

It can only be used with older vectors (pBR322) that carry two or more antibiotic resistance genes and an appropriate distribution of restriction enzyme cleavage sites. The foreign DNA is cloned in the plasmid in such a way that it disrupts the reading frame of one of the antibiotic resistant genes. Recombinant bacteria are screened by growing identical colonies separately on more than one antibiotic plate. If the bacteria become sensitive to the antibiotic, whose gene was disrupted by the insertion of the foreign DNA. and remains resistant to the others; it indicates that the bacteria contain the plasmid having foreign DNA.

3. Screening by Hybridisation

In this method the bacterial colonies are transferred on a nitro-cellulose paper and lysed to release and denature the DNA. that is immobilised on the nitrocellulose paper. This nitrocellulose paper is then hybridised with radioactive labelled DNA fragments that was used for cloning. The autoradiogram is aligned with the original bacterial-plate, from where the colonies were transferred, to identify the bacteria carrying foreign DNA.

4. cc-Complementation

Many of the vectors in current use (*e.g.*, the pUC series) carry a short segment of *E. coli* DNA that contains the regulatory sequences and the coding, information for the first 146 amino acids of the 13-ga1actosidase gene *(lac* Z). Embedded in this coding region is a polycloning or multiple cloning site that does not disrupt the reading frame but results in the harmless interpolation of a small number of amino acids in to the amino-terminal fragment of 13-ga1actosidase. Vectors of this type are used in host cells that code for the carboxy-terminal portion of the 13-galactosidase. Although neither the host-encoded nor the plasmid encoded fragments are themselves active, they can associate to form enzymatically active protein. This type of complementation, in which deletion mutants of the operator-proximal segment of the lac-Z gene are complemented by P-galactosidase-negative mutants that have the operator-proximal region intact, is called a-complementation. The lac^{k} bacteria that result from a-complementation are easily recognised because they form blue colonies in presence of chromogenic substrate 5- bromo-4-chloro-3-indoiy143-D-galactose (X-gal). However insertion of a fragment of foreign DNA into the polycloning site of the plasmid almost invariably results in production of an amino-terminal fragment that is not capable of a-complementation. Bacteria carrying recombinant plasmid therefore form white colonies. The development of this simple colour test has greatly simplified the identification of recombinants constructed in plasmid vectors of this type. It is easily possible to screen many thousands of colonies visually and to recognise colonies that carry putative recombinant plasmids. The structure of these plasmids is then verified by restriction analysis of mini preparation of plasmid DNA.

Requirements

1. LB agar plate containing ampicillin
2. IPTG and X-gal - To a premade LB agar plate containing ampicillin, add 40 111 of a stock solution of X-gal (20 mg/ml in dimethylformamide) and 4 JAI of a solution of IPTG (200mg/m1). Using a sterile glass spreader, spread the solution over the entire surface of the plate. Incubate the plate at 37°C until the fluid has disappeared. Because of the low volatility of

dimethylforrnamide, this can take up to 3 or 4 hours if the plate is freshly made.

Procedure

1. Inoculate the plate with 200 u.1 of transformation mix. Using a sterile glass spreader, spread the solution over the entire surface of the entire plate. After the inoculum has absorbed, incubate the plate in an inverted position for 12-16 hours at 37°C.
2. Store the plate at 4°C for several hours. This allows the blue colour to develop fully. Colonies that contain active 13-galactosida.se are pale blue in the centre and a dense blue at the periphery. White colonies occasionally show a faint blue spot in the centre, but these are colourless at the periphery.
3. Pick few blue and few white colonies for analysis by restriction endonuclease digestion.

11 Southern Blot Analysis

Southern hybridization technique is used essentially to identify and locate a specific DNA sequence among a pool of DNA fragments fractionated according to size by electrophoresis on a gel (usually agarose). The DNA to be analyzed is first digested with one or more given restriction enzyme(s), size fractionated on an appropriate gel, denatured, transferred from the gel to a solid support membrane (usually nitrocellulose or nylon) and immobilized on the membrane so that the relative position of different DNA fragments is preserved through transfer to the membrane. Detection of a specific DNA sequence in the pool of fractionated DNA is based on hybridisation of a labelled complementary DNA (or RNA) probe in solution with the immobilized denatured DNA on the nitrocellulose/nylon membrane (denatured DNA immobilized on a solid support cannot self reanneal but can hybridize with complementary single stranded nucleic acid molecules in solution). The formation of hybrid molecule(s) on the membrane is detected either by autoradiography (if the probe was radioactively labelled) or by appropriate calorimetric/chemiluminiscence reaction (if the probe was non radioactively labelled).

The conventional Southern blotting or Southern transfer, as developed by E.M. Southern in 1975, involves capillary transfer (or blotting) DNA molecules from the gel to an overlying nitrocellulose or nylon membrane. This is inexpensive, requires no special equipment or apparatus but takes a longer time (12-24h). In recent years, more rapid, methods of transfer have been developed involving either vacuum or electro-blotting. The vacuum blotting relies on suction pressure to force the nucleic acid molecules to move out of the gel matrix and get trapped on the underlying membrane. In electro-blotting, negatively charged nucleic acid molecules are driven out of the gel by electrophoresis onto the membrane lying next to the gel on anodic side.

The DNA molecules transferred to the membrane need to be cross linked to membrane so that they do not diffuse out during later stages.

Since the nitrocellulose or nylon membranes have high affinity for nucleic acids/proteins, it is essential that non-specific binding of the labelled probe to the membrane is completely blocked so that the labelled probe binds only when specifically hybridized with the complementary sequences in transferred DNA and there is no background signal. A simple gelatine coating of the blotted membrane that abolishes the need for the usual pre-hybridization step will be followed.

During hybridization the labelled probe is made available and is allowed to form stable Watson-Crick pairing with complementary sequences in the immobilized DNA. Following hybridization, a series of washing steps are employed to remove unbound or non-specifically bound probe molecules. The concentration of sodium salts determines the degree of mismatches permitted during hybridization between the immobilized DNA sequence(s) and the probe: lower the salt concentration, higher the stringency for accurate Watson-Crick pairing of complementary sequences.

The following protocol describes Southern hybridization following a simplified alkaline capillary blotting and using non-radioactive probe labelled with digoxigenin (employing DIG-dUTP as one of the precursor during random-primed synthesis); the resulting hybridization is detected using alkaline phosphatase conjUgated anti-digoxigenin antibody. The basic principle of this detection method is as follows: DIG-labelled dUTP is incorporated into the probe DNA by random-primed DNA synthesis (or nick translation): the DIG-labelled probe hybridizes to complementary DNA sequences in the blot and the anti-DIG-antibody binds immunologically to the DIG antibody is conjugated with alkaline phosphatase, when a suitable chrornogenic substrate for the alkaline phosphatase is provided, it results in appearance of a colored band at the site where the probe hybridized with the blotted DNA. Using this procedure, 0.1 ng or even less of plasmid DNA can be easily detected.

11.1. Restriction Digestion of Genomic and Plasmid DNA

Reactions typically contain 0.2 – I g of DNA in a volume of 20 pi However, reaction volumes may be increased or decreased according to need. Each restriction enzyme has a set of optimal reaction conditions, which are given on the information sheets of various catalogues. The major variables are incubation temperature and the composition of reaction buffer. Broadly buffers are of three types: high ionic strength, medium ionic strength and low ionic strength. Usually buffers are made up as 10X stock and then stored at -20°C.

Setting up of Digestion

1. Mix the following in a clean 1.5m1 Eppendorf tube kept on ice in the order given:

(i)	DNA (0.2 to 1.0 ug)	2 μl
(ii)	10X Restriction enzyme buffer	2 μl
(iii)	Restriction enzyme (units/pg DNA)	1 μl
(iv)	Distilled Water to make	20 μl

2. Mix the contents after each addition by tapping the tube (The actual amounts of each solution may be adjusted in relation to the amount of DNA being digested in the given case)

   (***Note***: A. The tube containing Restriction enzyme should be taken out of the -20°C freezer after the DNA and enzyme buffer have already been added *i.e.*, just before you are ready to add the enzyme; the enzyme should be replaced in the freezer immediately after addition of the enzyme to the reaction mix;

B. Volume of enzyme should not exceed one-tenth the total reaction volume because glycerol present in the storage buffer of enzyme may inhibit or alter the enzyme activity

3. Incubate the tube at appropriate temperature for the required period of time (one to a few hours).
4. Stop the reaction by heating the tube for 5 min at 75°C or by addition of 0.5M EDTA (pH 8.0) to give a final concentration of 10mM (*i.e.* 0.4 gl of 0.5M EDTA to a 20 il reaction volume).
5. Analyze the digested DNA on a gel.
6. When digesting bulk genomic DNA for Southern hybridization *etc.* (-10 [ig DNA per lane), restriction enzyme digestion is performed in a total reaction volume of 200 pl. After checking, a small aliquot for completion of digestion by running_ a test gel, the remaining of the DNA is ethanol precipitated, dried and redissolved in 15 to 20 p.1 of TE, mixed with sample buffer and run on agarose gel.

11.2. Agarose Gel Electophoresis of DNA

As per protocol V.

11.3. Southern Hybridization

Requirements

1. Capillary-blotting setup
2. Agarose gel with the desired DNA sample(s) separated by electrophoresis
3. Nylon membrane, Whatman filter paper (3 mm) and ordinary blotting papers
4. Polythene bags and bag sealer
5. Incubator/water bath set at 65°C for hybridization (preferable with a shaking or rotating device to ensure uniform distribution of hybridization solution in the bag)
6. Incubator set at 65°C (for washing)

Solutions Required

1. 20 X SSC (3 M sodium chloride, 0.3M sodium citrate)
2. 10 per cent SDS
3. Depruination solution (0.25 N HCl) I
4. Denaturation solution (0.5 N NaOH, 1.5 M NaC1)

5. Neutralisation solution (1 M Tris-HC1 (pH 7.5), 1.5 N NaCi)
6. 0.5 M EDTA (pH 8.0)
7. Digoxigenin-dUTP labelled DNA probe.
8. Colour detection buffers
    a. Buffer I (100mM Tris-HC1 (pH 7.5); 150mM NaC1)
    b. Buffer II (0.5 per cent w/v Blocking reagent in Buffer I)
    c. Buffer III (100mM Tris-HC1, (pH 9.5); 100mM NaC1, 50mM MgC12)
    d. Buffer IV (TE buffer (pH 8.0))
9. Nitroblue- tetrazolium (NBT)/5- bromo-4-chloro-3- indolyl phosphate (BCIP) solution
10. Hybridization solution:

    6X Denhardt's solution

    5X SSC

    25ng/m1 DIG labelled probe

Procedure

A. Capillary Blotting

1. Examine the ethidium bromide stained' gel under transilluminator to record (by photography or tracing on a transparent sheet) the positions of marker DNA bands and of any other bands of interest in the sample lanes.
2. Transfer the gel to a dish and trim away all the unwanted areas.

    Note: Handle the gel with gloved hands since ethidium bromide is a potential carcinogen.
3. **Depurination:** Add 0.25 M HCl (enough to submerge the gel) and leave for 8 to 10 min with intermittent shaking (the bromophenol blue dye front will turn from blue to greenish yellow)
4. **Denaturation:** Rinse the gel once in distilled water and submerge in denaturation solution for 15 min with constant, gentle agitation to denature DNA (the bromophenol dye regains its blue color).
5. **Neutralisation:** Rinse the gel once in distilled water -and submerge in neutralisation solution with constant agitation for 30 min.
6. Set the 2e1 for capillary blotting:
    - ☆ Place a 3mm Whatman filter sheet on the blotting apparatus followed by a polythene sheet in which a window of exact size of the gel has been cut.
    - ☆ Soak a nylon membrane, cut to the size of-the gel in 2X SSC in a dish and letting it soak uniformly by itself.

- ☆ Position the gel with denatured DNA (wells facing down) in the window in polythene sheet and carefully align the of the nylon membrane on the gel followed by a 3-4 sheets of Whatman 3 No. sheets and a stack of filtet papers. Make a small cut on one corner of the membrane to provide a reference point with respect to the gel for later analysis.
- ☆ A membrane that does not soak uniformly within 1 or2min may not be suitable for transfer.
- ☆ Do not touch the nylon membrane *etc.* with bare hands since greasy marks spoil the membrane: always use gloves while setting up blotting.
- ☆ It is important that size of the window in the polythene sheet and of the nylon membrane and filter papers in the top stack exactly matches the gel size.
- ☆ Place a weight of approximately 500 ern over the stack of filter papers and leave the assembly for 8-24 hours by chancing the filter papers

  ***Note*:** Extreme care has to be taken that no air bubbles are trapped between any layers of the entire stack: any trapped bubbles are immediately removed by gently rolling a pipette on the stack.

7. After this period, the stack of filter papers would have almost wet the agarose gel itself would have become very thin. The bromophenol dye can be seen on the nylon membrane
8. After the transfer, wash the membrane in 2X SSC and let it dry at room temperature
9. Crosslink the transferred DNA with the nylon membrane by exposing the DNA side of the filter to UV for 4 min on a transilluminator (if using nitrocellulose membrane, crosslinking is done by baking the dried membrane at 70°C in vacuum) (Alkaline transfer by itself is known to crosslink DNA with nylon membranes; therefore, the UV crosslinlcing is not absolutely essential).

B. *Labelling of Probe*

1. Prepare the labelling mix in the following order:

Denatured DNA	- 241 (– 1 pg)
Hexanucleotide mix	- 2111
dNTP mix	- 2111
Klenow enzyme	- 1111
Distilled water to	- 20 tit

Incubate the mixture at 37°C for 1 – 20 hours in water bath.

2. To precipitate the labelled DNS, ad 0.1 vol of 4M LiC1 and 2.5 vol of chilled ethanol and keep at -70°C for 2 hours or at -20°C for overnight.
3. Quantitate the labelled DNA against the control labelled DNA.

C. *Hybridization*

1. Add prehybridisation solution to the bag containing the blotted membrane. Carefully seal the bag with bag sealer after removing as many air bubbles and incubate at 65 °C for 2 – 6 hours.
2. Denature the probe by placing the tube containing the desired amount of labelled probe in a boiling water bath for 5 – 10 min and immediately chill on ice. Add the denatured probe (25 ng/ml) to freshly prepared prehybridization solution mix and pour the entire amount into the bag with membrane. Seal the bag and incubate at required temperature for 8 – 24 hours.

D. *Post Hybridisation Washing*

1. After hybridization, take the filter out of the polythene bag into a small dish and add 2X SSC, 0.1 per cent SDS enough to submerge the membrane. Keep for 10min at room temperature.
2. Transfer the membrane to another dish containing 0.1X SSC, 0.1 per cent SDS at 64°C. After 15 min, remove the solution and add fresh 0.1X SSC, 0.1 per cent SDS and keep at 64°C for another 15min.

   Note: The temperature and concentration of SSC during washing determines the stringency of hybridization; lower temperature and/ or higher SSC concentration reduce stringency so that poorly matched hybrids also stay. This is useful especially when probing with a heterologous probe which may have reduced sequence homology with the target DNA on the blot.

E. *Colorimetric Detection of Hybridization*

1. Wash the filter briefly in buffer **I.**
2. Incubate in Buffer II for 30min at room temp (to block the membrane surface for non-specific binding of the antibody used in next step).
3. Briefly rinse in Buffer I.
4. Incubate in Anti-DIG Antibody -Enzyme conjugate (1 p.1 in 4 ml of Buffer I) for 30min at room temp
5. Wash twice at 15 min interval in Buffer **I.**
6. Briefly rinse in Buffer **III.**
7. Put the blot in a small polythene bag and working in dim light add color developing solution NBT and BCIP solution in 1 – 5 ml of Buffer III) and seal the bag.

8. Incubate the blot within the sealed bag in dark (*e.g.*, by wrapping with aluminium foil) till desired level of colored signal is visible (a few minutes to several hours depending upon the amount of target DNA on the blot, sequence homology and efficiency of DIG- labelling of the probe).
9. When adequate signal is obtained, remove the blot from the bag and put in Buffer IV or in dry condition.

If reprobin2 with another probe is desired, the blot should not be dried

The color is photosensitive and may fade with time; original color may be restored by wetting the blot once again with Buffer IV. When stored in Buffer IV, some background color may also appear With time.

12 Immunocytochemistry

In immunocytochemistry an antibody is used to link a cellular antigen specifically to a stain that can be readily seen with a microscope. Detection of antigens in tissues is known as immunohistochemistry, while detection in cultured cells is generally termed immunocytochemistry. We have a wide range of specimen source, antigen availability, antigen antibody affinity, antibody type, and detection enhancement methods from which selections can be made for immunohistochemical or immunocytochemical detection optimally dependent on the above variables. As for all procedures, reference should be made to individual product data sheets and published literature. Also, the Internet today contains a tremendous amount of useful information on Immunohistochemistry/immunocytochernistry. Quick searches through google.com or similar search engines are highly recommended.

Specimen Preparation

Fixatives

To stop post mortal changes and/or to preserve cells and tissues in a reproducible and life-like manner one has to fix the specimen. To achieve this, tissue blocks, tissue sections, cell growths or smears are either immersed in a fixative fluid. In cases of whole animal systems, the animal is perfused with fixative via its circulatory system. In the case of cells in culture. cell preparations are either submerged or simply air-dried. Fixation procedures stabilize cells and tissues and thus protect them from the rigors of subsequent processing and staining with solvents and dyes. For immunological studies, one must ensure the adequacy of the specimen and target antigens. Thus the choice of fixative and the processing' are crucial. Tissues have differing protein content and structural arrangement, thus they have a variable ability to retain their structure with or, without fixation. Incorrect specimen preparation can block or impede antigen labeling in tissue and cells. Unfortunately, the methods that are best for the preservation of tissue structure do so by altering proteins, thereby masking some epitopes, and may even prevent the detection of the desired target protein. Hence standardization of an optimal procedure and selection of an appropriate fixative may be needed for greater success.

Fixatives may work by several means: formation of crosslinkages (additives,*e.g.*, aldehydes such as glutaraldehyde or formalin); protein denaturation by coagulation (coagulants, *e.g.* acetone and methanol); or a combination of the above. Fixation strengths and times must be optimized so that antigens and cellular structures can be retained and epitope masking is minimal. Requirements for fixation can vary widely between tissues. For example when using antibodies to probe for neurotransmitter substances, most tissues must he either immersion fixed with a mixture of glutaraldehyde and paraformaldehycle, or with paraformaldehyde alone.

Immunocytochemistry

Fixation of cultured cells for immunocytochemistry also requires careful consideration. Fixation strengths and fixation times are considerably shorter for cells as compared to the thicker. tissue sections. For immunocytochemistry, sample preparation essentially entails fixing the target cells to the slide. Depending upon the needs of the experiment, cells can be simply harvested and "dropped" on to a slide, fixed and air-dried, as is often done with clinical or diagnostic studies where speed and a simple "yes/no" answer is all that is needed. Alternatively, where information on the structural location within the cell is needed, cells are often cultured directly on prepared slides or coverslip material, which are then simply washed and fixed in place prior to staining. Consulting published literature relating to the tissue/ proteins of interest is well worth the time invested. Recipes for common fixatives are available in most histology texts.

Section Type

It is important to decide which type of sections is to be used. Fixed or unfixed cryostat (frozen) sections. fixed "wet" or vibratome sections, or fixed, paraffin-embedded sections can be used. The choice of section is determined by a number of issues, including the time and skill of the investigator. While fixed frozen sections are often quickest and easiest to use. because of their superior fidelity, clarity, and preservation properties. fixed paraffin-embedded tissues have become the ultimate standard of immunohistochemistry in histology and pathology.

Cryostat (Frozen) Sections

There are two types of Cryostat sections: (1) Fresh, frozen, snap frozen tissues are first cut, then either air-dried or fixed prior to staining; (2) or fixed frozen sections. where 'the tissue is first liked then cryoprotected prior to freezing and then sectioned. The advantages of frozen sections are that they allow excellent antigen preservation, they are typically faster to perform, and they offer flexibility, since any fixative can be used, thereby facilitating the optimization of fixative for each antigen. However frozen sections give less morphological detail and resolution than paraffin sections.

Sample Protocols for Cryostat Sections - Fresh Frozen (then fixed) Tissue Sections

1. Snap-freeze small tissue blocks (5x5x3 mm) in liquid nitrogen.

2. Transfer to cryostat and cut thin (5-30 um) sections.
3. Collect specimens on clean poly-L-lysine/chromalum gelatin - coated glass slides and dry at room temperature overnight (if you want to stain the same day let air-dry for 1-2 his. until completely dry). Thorough drying is essential for good sections adhesion to the slides.
4. Fix sections in acetone or absolute ethanol at 4°C for 15 min. Use fresh ethanol or acetone for every 10-15 slides for best results.
5. Thoroughly air-dt y at room temperature or on mild heat (30-37°C).
6. Proceed with immunostaining or freeze.

Fixed, Frozen Tissue Sections

1. Fix tissue either by perfusion with fixative or by immersion in fixative (*e.g.*, 4 per cent paraformaldehyde for a set time period.
2. Fixed tissue is then cryoprotected by submerging the target tissue in 10, 20, 30 per cent sucrose solution in PBS till the tissue sinks.
3. Once stabilized, tissues can be removed from the protectant solution and frozen at -70°C until sectioned.
4. Sectioned via cyrostat (5-40 p.m), where sections can be collected directly onto slides, or floated onto slides via a PBS/waterbath. Usually up to 3 sections per slide can be placed; each spaced well apart. The spacing prevents reagent mixing between samples. Individual skill and tissue type determine the thickness of the sections. Sections between 10-15 pm provide the best results for clarity and integrity. Sections between 6-9 p.m tend to tear during cutting, resulting in rough edges that can increase the background. Thicker sections while stronger during handling are usually difficult to stain.
5. Sections on slides are thoroughly air/warmed dried on a slide warmer, usually overnight or at least 2-3 hrs. at 40-50°C.
6. Prepared slides can be stored dry at -70°C until stained. Equilibrate to room temperature and briefly redry prior to rehydration and staining.

Paraffin Sections

The largest proportion of samples used in immunostaining are embedded in paraffin because it provides for excellent morphological detail and resolution. Modern "paraffin" is typically a mixture of paraffin wax and resin. It is an excellent embedding medium because it can be heated to liquid state, and dissolved by toluene/xylene for infiltrating the tissue, and then relatively quickly turned to a solid state again for maximum structural support during sectioning. Typically, small blocks (10x10x3 mm) of tissue are fixed for up to 24 hrs. The most common fixatives used in paraffin sections are formalin-based. Recipes of common fixatives are listed at the end of this protocol. The blocks are then infiltrated and embedded with paraffin and 5-10 pm sections are cut in ribbons and mounted on slides. Once mounted, the slides can be stored indefinitely until immunostaining is required.

Sample Protocol for Paraffin-embedded Sections

A. Conventional deparafftnization and rehydration sequence:
   1. Deparaffinise sections in Xylene: 2 to 3 changes, 5 min. each.
   2. 100 per cent absolute ethanol: 2 changes, 3 min. each
   3. 95 per cent ethanol: 2 changes, 3 min. each
   4. 80 per cent ethanol: 3 min.
   5. 50 per cent ethanol: 3 min.
   6. Rinse with distilled water, PBS, or Tris buffer: 2 changes. 3 min. each.
   7. Without allowing the sections to -dry follow the following steps.

B. Place slides in prewarrned (37°C) 0.1 per cent trypsin in PBS for 5-60 min. or 0.4 per cent pepsin in 0.01N HC1 for 30 min. to one hour. Follow by rinsing with distilled water.

C. If peroxidase conjugate is used endogenous peroxidase should be blocked at this point. Peroxidase activity results in the decomposition of hydrogen peroxide (1-1202). Suppression of endogenous peroxidase activity in formalin-fixed tissue incubated sections in 3 per cent 14702 for 8-10 min. Alternatively methanolic 1-1202 treatment (1 part 3 per cent H_2O_2 plus 4 parts absolute methanol) for 20 min., but it is not recommended for specimens where cell surface markers are to be stained.

D. Wash twice with PBS

E. Proceed with immunostaining procedure (see Antibody Staining section).
   Antigen Retrieval

To facilitate the immunological reaction of antibodies with antigens in fixed tissue, it may be necessary to unmask or "retrieve" the antigens through pretreatment of the specimens. Different antigens and different antibodies will require different antigen retrieval methods. Antigen retrieval has been shown to increase reactivity of the majority of antigens in tissues.

Antigen retrieval includes a variety of methods by which the availability of the antigen for interaction with a specific antibody is maximized. The most common techniques are enzymatic digestion or heat induced epitope retrieval (1-HER) through microwave irradiation, autoclaving or pressure cooking.

Enzymatic Digestion

Rehydrated sections are equilibrated with the appropriate buffer, and incubated with a proteolytic enzyme at 37°C, or at room temperature. Enzymes used include protease (0.05 per cent (w/v) in PBS), trypsin (0.05 per cent (w/v) in PBS with 0.1 per cent CaC12) and pepsin (0.05 per cent (w/v) in 2 N HCl). The concentration, time and temperature must be controlled, so that the enzymes can break some of the bonds formed during fixation. uncovering antigenic sites. but the antigen should not be digested completely. The enzymatic activity is stopped by placing the specimen in cold buffer (4°C) prior to processing with antibody. These methods should be selectively used.

Microwave Irradiation

Microwave irradiation of formalin-fixed, paraffin-embedded specimens in buffer has been found to be effectively used in retrieval of antigens. This procedure helps break some of the bonds formed during fixation, thus increasing the number of positive cells available, and the intensity of reactions. It is important to monitor the sections during the microwaving process, to prevent damage and drying. To avoid variability the buffer volumes, irradiation times,, and microwave unit used, will must be kept constant during the entire course, of the process. Periodically, the heating is 'stopped and liquid is replenished. After a set time, the solution containing the slides is allowed to cool to room temperature slowly followed by rinsing in PBS and then used for staining.

Autoclaving or Pressure Cooking

It is useful if the procedure is standardized. It is important to start with standard volumes of preheated solutions. After adding the specimens to the boiling retrieval solution, the autoclave or pressure cooker should be brought to full pressure as quickly as possible and the heating times measured exactly from this point. At the end of the heating time (usually 1 to 2 minutes) the pressure should be released and the hot buffer should be quickly flushed out with cold water. Without drying up the sections. the specimens should then be washed in buffer at room temperature.

A simple incubation in citric acid buffer, pH 3.0 (2.1 grams Citric Acid added to 400 mL of ddH20 Adjust to pH 3.0 with Acetic acid if above 3.0, or NaOH if below 3.0, make up to 1 L final volume with ddH20) for 30 minutes, at 37°C after blocking but prior to primary antibody addition could be a more mild procedure. Rinse slide in PBS or TBS, pH 7.4 prior to staining.

Antibody Staining

Primary antibody may be directly labeled with an enzyme (such as horseradish peroxidase or alkaline phosphatase) or fluorophore (such as FITC or rhodamine), or unlabeled, with detection by a labeled secondary antibody or more complex detection system. If a secondary antibody is used, it must be generated against the immunoglobulins of the primary antibody source, *e.g.*, if the primary antibody is raised in rabbit, then the secondary antibody could be goat anti-rabbit. The optimal titer of both the primary and secondary antiboay should be determined for each batch.

The proper working dilutions for every antibody must be optimized for the system in which it is being employed. The same system does not always work for every antibody. The product data sheets may be used as a guide for dilution series starting points. The optimal antibody dilution will be the one that gives the strongest specific antigen staining with the lowest non-specific background. As with other controlled experiments, it is advisable to change only one experimental variable at a time. After determining the optimum titer/dilution of the primary antibody, the secondary antibody dilution can be optimized.

For staining of tissue sections, it is customary to incubate with 25-50 i.t1 of diluted antibody – the volume used must be sufficient to completely cover the tissue, and to ensure the tissue will not dry out during incubation. Incubation times may range from 30– 90 minutes at 37°C, from one to six hours at room temperature. or overnight at 4°C. Incubation times should he optimized empirically for each antibody'antigen combination.

General Protocol for Immunohistochemical Staining with Polyclonal Rabbit or Monoclonal Mouse Primary Antibody and Employing Streptavidin Biotin HRP Method Imunofluorescence Method

The following, general protocol is intended for use as a guideline in developing antibody-specific procedures. Different antibodies and tissues may require changes to this procedure. Review of individual product datasheets and relevant literature references may be helpful in customizing this procedure for specific applications.

1. Gently rinse slide containing sections with buffer either PBS or TBS. Place slide in room temperature buffer bath for 5 minutes to rehydrate sections.
2. Perform endogenous peroxidase blocking, if using peroxidase system by incubating sections in 3 per cent H202 in PBS for 20-30 minutes followed by washing in 3 changes of PBS for 5 minutes each. Omit this step if using immunofluorescence method.
3. Treat the sections with 0.2-0.5 per cent Triton-X100 in PBS for 20-30 minutes or perform any other antigen retrieval method opted for the primary antibody to be used followed by washing in 3 changes of PBS for 5 minutes each.
4. Non specific proteins can be blocked by using 1 per cent normal serum (normal serum from the host of the secondary antibody) or 1 per cent BSA or non-fat dried milk powder. Incubate for 40 minutes at room temperature.
5. Tap off serum and wipe away excess. Do not rinse.
6. Apply 100-200 pi, of primary antibody (monoclonalipolyclonal), diluted in 1 per cent BSA in buffer per tissue section. Antibody should cover sections 'completely. Incubate the sections at 4°C overnight. If optimal antibody dilution is unknown, perform a series of antibody dilutions in

the range of 1:20-1:1,000 or even more in case of polyclonal antibodies to obtain best results.

7. Rinse slide gently with buffer for 3 x 5 minutes (changing buffer in between washes).
8. S. Apply secondary antibody biotin conjugated for SABC method/ fluorochrome conjugated for IF method. The antibody should directed against the host in which the. primary antibody is raised. The secondary antibody may be diluted appropriately in 1 per cent BSA in PBS and the sections may be incubated for 45-60 minutes at room temperature.
9. Rinse slide gently with buffer for 3 x 5 minutes (changing buffer in between washes).
10. For IF studies the sections may be mounted at this stage in an anti-fade aqueous mounting medium from Sigma and can be visualized with fluorescence microscope.
11. For SABC method incubate the sections with streptavidin biotin HRP reagent after appropriately diluting in 1 per cent BSA in PBS and the sections may be incubated for 45– 60 minutes at room temperature.
12. Rinse slide gently with buffer for 3 x 5 minutes (changing buffer in between washes).
13. Incubate sections with substrate-chrornogen solution: DAB (3'3 Diamino benzidine from Sigma) 25mg/100 ml of buffer and 60p1 of H_2O_2 and incubate until desired color intensity has developcl.
14. Rinse gently with distilled water Counterstain with hematoxylin or any other nuclear stain, air dry and mount in entellan or alternatively dehydrate and mount in DPX.

While undertaking the hands-on training we shall describe/discuss several of the precautions one has to observe and several issues on the troubleshooting. Several theoretical aspects shall be covered by the authors in their technical lectures.

Antibody Detection

- ☆ Enzyme-Mediated
- ☆ Fluorescence
- ☆ Signal Amplification

Two of the most commonly used detection methods are fluorescence and colorimetric (enzyme mediated) detection. With the advent of electron microscopy, detection of antigens by antibodies that contain large gold particles is often used. and these may also be visualized at the light microscopic level as well, but their use is quite rare today, outside of electron microscopy. Described below are the common antibody detection methods for light microscopy.

Enzyme-Mediated Detection

When choosing a substrate for conversion by an enzyme, one should select a substrate which yields a precipitating product. Examples of commonly-used substrates are listed below.

Enzyme Unrseradish Broxidase			
Substrate	Observation	Emal Colour	Sodium
Diaminobendidine	DAB	Brown	No
Comments: 3,3'- diaminobenzidine produces a brown end product which is highly insoluble in alcohol and other organic solvents. Oxidation of DAB also causes polymerization. DAB has the ability to react with osmium tetroxide, and thus is very useful in electomicroscopy as well as traditional immunihistochemistry sections.			
Diaminobenzidine w/nickel enhancement	DAB/Nickel	Gray/Black	No
Comments: Produces a more intense stain which is resistant to alcohol and provides better contrast, up to 40 times more senstive than DAB wihtout enhancement.			
3- Amino 9- ethylcarbazole	AEC	Red/Brown	Yes
Comments: AEC produces a red/brown reaction product and is widely used for immunohistocehmical staining. Slide specimens processed with AEC must no be immersed in alcohol or alcoholic solutions (*e.g.*, Harris' hematoxylin). Instead. an aqueas couterstain and mounting medium should be used. AEC is also susceptible to further oxidation when exposed to light and thus it will fade overtime. Dark storage and brief light viewing are recommended			
4- Chloro-1-naphtol.	4-CN	Blue/Gray	Yes
Comments: 4- Chloro-1- naphtol (CN) precipitates as a blue end product, Because CN is soluble in alcohol and other organic solvents, the slides must not be dehydrated. exposed to alcoholic counterstrains or coverslipped with mounting media containing organic solvents. Unlike DAB, CN tends to diffuse from the site of precipitation, thus it is not usually recommended for Immunohistochemistry but can be used for Wetern blotting.			
Enzyme Alkane Phosphate			
Substrate	observation	Emal colour	S
Naphthol-AS-B1-Phosphate/fast red TR.			
Comments: Napthol AS- Acts as the substrate for alkaline phosphates, and the Fast Red chromogen precipitates at the enzymatic sites producing a vibrant red/pink color. Precipitate is soluble in in alcohol, thus aqueous counter stain and mounting medium should be used.			
Napthol-AS-MX-phosphate/fast red TR.	NAMP/FR	Red	Yes
Comments: Napthol AS-acts as the substrate for alkaline phosphates, and the Fast Red chromogen precipitates at the enzymatic sites producing a vibrant red/pink color. Precipitate is solvable in alcohol, thus aqueous counter stain and mounting medium should be used.			
Napthol-AS-MX-phosphate/new fuschin.	NAMP/NF	Red/Violte	Yes
Comments: Napthol AS-acts as the substrate for alkaline phosphatase, and the new Fuchsin chromogen precipitates at the enzymatic sites producing a vibrant red/violet color. Precipitate is soluable in alcohol, thus aqueous counter stain and mounting medium should be used.			

Bromochloroindolylphosph2te/Nitro Blue Terazolium.	BCIP/NBT	Purple	No
Nitro Blue Tetrazolium *Comments*: 5- bromo,4-chloro,3- indoylphosphate (BCIP)/nitroblue tetrazolium (NBT) substrate is a commonly used substrate chromogen. BCIP acts as the substrate for alkaline phospatase, and the NBT enhances the purplish-brown color of the precipitate. BCIP/NBT is compatible with organic solvents so it can be used with alcohol based counter stains including Nuclear Fast Red or Methylene-Green			

Fluorescence

A molecule that fluoresces can be attached to the antibody for detection using UV light. Examples are Fluorescein, Rhodamine. Texas Red®, Cy3 and Cy5. In selecting fluorochromes, one is limited by the available microscope filter sets. Most filter sets are best matched with rhodamine or fluorescein. Texas Red may also be used with a rhodamine filter set.

Many mounting. media contain "anti-fading," solutions, such as DABCO, which will prolong the viewing time of the sample. Chemicon offers a variety of fluorescence mounting fluids and counterstain solutions including our basic fluorescent mounting fluid (Cat. No. 5013) and enhanced counterstaining fluid containing nuclear stains such as DAN and PI (Cat. No. S7113, S7114).

Texas Ree and Alexa Fluor® are registered trademarks of Molecular Probes, Inc. Fluoro-Jade® is a registered trademark of Histo-Chem, Inc.

Signal Amplification

Signal amplification techniques greatly enhance the sensitivity of immunohistochemical and immunocytochemical methods. The signal amplification methods may be used in conjugation with either of the above detection techniques. Signals may be amplified by using poly-conjugated secondary antibodies (*i.e.* Chemicon catalog numbers AP340P–AP342A series). or Avidin-Biotin interactions or other commercially-available amplifiers (i.e tyramide catalyzed systems). which increase the signal to antibody ratio. When signal amplification is used to amplify the specific signal. however, one should be aware that non-specific signals may also become amplified. Thorough washing and proper antibody 'titration is espetially important in this case.

Troubleshooting

- ☆ No Staining
- ☆ Weak Staining
- ☆ Background Staining

When tissue staining has not given the expected results, the experiment should be examined in a systematic way, wherein only single experimental variables are altered at one time. Proper immunohistochemical troubleshooting requires one to determine whether difficulties are related to specimen, antibodies, technique, environment, or slide interpretation? The following checklist may assist in troubleshooting efforts.

No Staining of Either Controls or Specimen

- ☆ Confirm that no reagents were omitted (primary antibody, secondary antibody, substrate components)
- ☆ Confirm that reagents were added in the correct order, and for sufficient incubation times.
- ☆ Re-read labels to confirm that correct antibodies were used. This is especially important when using primary/secondary antibody combinations. For example, when using a mouse IgM primary antibody, the secondary antibody should be a goat or rabbit anti-mouse IgM (not IgG).
- ☆ Check antibody titrations and dilutions. This is particularly important for the primary antibody.
- ☆ Check that antibody blocking solutions or diluents do not contain substances that may interfere with or absorb the primary antibody. For instance, serum containing diluents can absorb out or reduce primary reactivity to small molecules such as GABA. or amino acids.
- ☆ Check reagent expiration dates and storage. Enzymes and fluorochromes are especially prone to breakdown after prolonged storage. Antibodies should be stored in nondefrosting freezers, as self-defrosting models will expose antibodies to repeated freeze/thaw, resulting in antibody breakdown.
- ☆ Check specimen storage. Where possible, compare staining of the unknown specimen versus a known positive tissue in a "side-by-side" experiment.
- ☆ Check the chromogen/substrate solution. This can be checked by adding a drop of the labeling reagent to a small sample of prepared chromogen. If the chromogen is working, the mixture should change color. Chromogen solutions can deteriorate quickly, so do not use beyond the times recommended by the manufacturer. Lack of color change may be due to inactive enzyme or improperly prepared chromogen.
- ☆ The rinse buffer may be incompatible with the reaction reagents. The pH must be appropriate, and buffers to be used with peroxidase enzyme should not contain NaN_3.
- ☆ The counterstain and mounting media may not be compatible with the chromixten. Check the manufacturer's recommendations.
- ☆ Check that the microscope is adjusted correctly and that the fluorescence lamp is not burned out.

Weak Staining

Points to consider are:

1. Is the intensity of the staining consistent between the positive controls and the test sample(s)?
2. Is the staining specific for the antigen of interest, or is it background staining? This can only be determined by examining the slides.

All the items listed above for No Staining can apply to a lesser degree to the situation of weak staining. However, if the negative controls are devoid of stain and the positive controls and test sample(s) are weakly stained, then possible trouble points include:

- ☆ Overfixation, or incorrect fixation for the immunological procedure in use.
- ☆ Insufficient antigen retrieval
- ☆ Antibody concentration may be too dilute: If possible, the concentration should be increased. If this is not feasible, then the incubation time or temperature may need adjusting. When diluted antibody is stored in the refrigerator it sometimes gets absorbed to the walls of the container. Storing the antibody with a protein carrier such as 1 per cent –3 per cent BSA can alleviate this situation.
- ☆ Too much buffer rinse has been left on the slide, so that the antibody becomes diluted when added to the sample.

If the negative controls have not reacted, the positive controls are well stained, but the test sample is stained weakly, then either the positive control and the test sample were fixed differently, are of different tissue type, or the outer tissue of the test specimen block has been poorly fixed.

If the negative controls have not reacted, the test sample(s) are well stained, but the controls are weakly stained, then the control material should be replaced.

Background Staining

If the negative control is being stained as well as the positive controls and test sample(s), then the degree and type of background staining must be analyzed. The following are possibilities for investigation:

- ☆ Re-titer antibodies (both primary and secondary) with a dilution series.
- ☆ Incubate with chromogen for a shorter time. Some chromogens, such as DAB, develop very quickly.
- ☆ The chromogen was not totally dissolved, and associated with the tissue. Centrifuge or filter the chromogen solution.
- ☆ Particulates in the antibody solution. These may form upon repeated freeze/thaw and can be eliminated by centrifugation.

- ☆ Insufficient rinsing between steps, or contaminated buffers. Mix new buffers and increase washing steps.
- ☆ Enzyme or biotin in the tissue is reacting with the reagent. This can be prevented by increasing the time or concentration of block, trying different types' of block, or using a combination of more than one block, or changing the staining methods. Some tissues (*i.e.* brain and liver) are known to contain high endogenous levels of biotin or peroxidase activity.
- ☆ The incorrect blocking serum was used, or blocking serum was not used. The blocking serum should be from the species of the secondary antibody. It is possible to use 5 per cent nonfat dry milk rather than serum.
- ☆ The secondary antibody cross-reacts with endogenous tissue proteins. Secondary antibodies which have been absorbed against immunoglobulin from the species from which the target tissue was obtained will result in significantly lower background, and are indespensible for double-labeling experiments.
- ☆ Hydrophobic and/or ionic interactions between the reagents and tissue types such as connective, adipose or fatty tissues may give rise to apparent specific reactions. Antigen retrieval procedures can be of great assistance in correcting this predicament. A decreased fixation time in formalin can also help.
- ☆ The embedding media may not be completely removed from the tissue. Review the removal procedure for possible changes.
- ☆ The specimen may have dried out during the procedure, allowing the trapping of reagents under the edges of the specimen. Care should be taken to avoid letting the specimens dry.

If there is background staining in the positive controls and the test sample(s), but not, in the negative control, then the issue is most likely associated with the primary antibody. Some possibilities are:

- ☆ The primary antibody was too concentrated. or the incubation period too long. More dilute antibody, or shorter incubation or lower incubation temperature may correct the situation.
- ☆ The tissue may contain Fe receptors, or there may be interfering Ig components (aggregates or oligomers) or there may by naturally occurring, contaminating antibodies. This can be resolved by using Fab fragments rather than whole IgG molecules, filtering out the aggregates, or by diluting the primary antibody and incubating for longer times.
- ☆ The tissue sections may be cut too thick – try thinner sections.
- ☆ The microscope light needs to be adjusted to a higher setting.
- ☆ If there is background staining in only the test sample(s) - *i.e.* not in the positive or negative controls, then the most likely cause is that the test sample(s) has been fixed and processed differently from the controls.

Use of different tissue type between test sample(s) and controls may also produce this variance. Possibilities include:

- ☆ Overfixation of the test sample, resulting in the increased presence of hydrophobic groups. or increased crosslinking. Use of antigen retrieval procedures will amend this.
- ☆ A different fixative was used for the test sample(s) than for the control tissue.-This difference should be avoided, or. the procedures should be adjusted.
- ☆ The test sample(s) and the controls are of different tissue type. This should be avoided whenever possible.

If the test sample(s) and positive controls are clean, but the negative control shows background staining, it is likely that the negative control serum is at fault. It may be too concentrated, or contaminated with cross-reacting Ig components, naturally occurring antibodies, or bacterial growth. This can be corrected by using more dilute serum and incubating longer, trying to find a better match for the negative serum. or purifying, the serum.

13 Western Blotting

Polyacrylamide gels as solid supports for electrophoresis were introduced by Raymond and Weintraub in 1959. Weber and Osborn (1969) introduced sodium dodecyl sulphate (SDS) treatment of protein before their electrophoresis, which facilitat63 separation of polypeptides purely according to the differences in their molecular weights. In 1970 Laemli improved and modified this technique and introduced slab gels in place of tubes. The Laemli method of SDS-PAGE is extremely popular and is commonly used by the most of the protein research laboratories.

Sodlunj clodecyl_sulphate, - polyacrylamide gel electrophoresis is a technique often employed for,leparation_of large number,ofpro-teins in heterogeneous, samples largely on the basis of their molecular weighti, Accurate molecular weight determination in this procedure depends upon two factors:

1. Proteins in general bind with constant amounts of SDS per gram when saturated. The protein then has an overall negative charge that masks its intrinsic charge resulting in a constant charge to mass ratio for proteins
2. Proteins saturated with SDS take on a rod like configuration, the length of the structure being proportional to the polypeptide chain length and thus the "molecular weight. SDS-PAGE has also got the advantage of separating the subunit components of a protein.

13.1. Sample Preparation for SDS-PAGE

Requirements

(a) **Instruments, ware and others:** Dissection trays, scissors, ice bucket, dissection microscope and dissection needles (for fly Models), chloroform or diethyl ether (anaesthetic agent)

(b) **Reagents:**

1. lox pm (Dissolve 7.6 gm NaCl. 1.246 gm Na_2E1PO_4. 0.468 gm NaH_2PO_4 in 100 ml distilled water. Adjust pH between 7.0 and 7.4)
2. 2X sample/lysis buffer (50 mM Tris-HCl (pH 6.8) 2 per cent SDS, 10 per cent glycerol)
3. IX sample/lysis buffer (50 per cent 2X sample buffer. 5 per cent 13-mercaptoethanol)

Procedure

1. Dissect out the tissues (*e.g.*, Liver, Kidney. Blood cells or Serum) from model organism, rinse in 1X PBS and immediately transfer on ice in aluminium foil.
2. Take the small tissue piece (approximately 25 mg), mince on ice and transfer in 200 l 1X sample buffer and store at -20°C till use.
3. Denature the proteins in boiling water bath for 3-5 minute and immediately transfer on to ice and centrifuge at 10,000 rpm at 4°C for 10 minutes.
4. Collect the supernatant in a separate microfuge tube and again centrifuge for 5 minute to settle the small tissues or debris.
5. Transfer the samples on ice till estimation and loading on slab gel.

13.2. Estimation of Protein by Modified Bradford Method

Requirements

(a) Glassware, instruments and others: Test tubes, pipettes (P-20, P-200 and P-1000), pipette tips, test tube stand, spectrophotometer

(b) Reagents:

1. Bovine serum albumin (10 mg/ml)
2. 10 per cent Trichloro acetic acid (500 gm TCA in 227 ml makes 100 per cent w/v TCA)
3. 0.15 N NaC1
4. 1M NaOH
5. CBB G-250 dye

Procedure

Preparation of Standard Curve of Bovine Serum Albumin (BSA)

1. Take the 2, 4 6, 8, 10 12 and 14 p.1 of 10 mg/ml BSA in separate test tubes in triplicates.
2. Add 3 µl of 10 per cent Trichloroacetic acid for precipitation of protein. Add equal volume of (volume of BSA + volume of TCA) 1 M sodium hydroxide.
3. Make up the total volume to 100 ul with 0.15 M sodium chloride.
4. Add 1 ml of distilled water and 1 ml of CBB-G 250 dye.
5. Take the OD at A_{595} and plot the standard curve on graph paper with concentration on X-axis and OD on Y-axis.

Protein Estimation of Tissue Samples

1. After denaturation step (13.1 step 5), take 2μl of protein sample in separate test tube and add 3 μl of 10 per cent TCA for precipitation.
2. Add 5 μl of 1M Sodium hydroxide and 90 μl of 0.15 M sodium chloride.
3. Add 1 ml distilled water and I ml CBB-G 250 dye.
4. Take the OD at A_{595} arid calculate the protein amount in 2μl volume of sample with the help of standard BSA graph

13.3. SDS-PAGE

Requirements

(a) Glass ware and others: Gel plates (notched and rectangular), microfuge tubes, micropestles, glass/plastic tray, staining boxes

(b) Instruments: Gel apparatus with power supply, platform shaker

(c) Reagents:

1. 30 per cent Acrylamide (29.2 gm acrylamide, 0.8 gm N,N-biS acrylamide).
2. Separating get buffer (1M Tris-HCl (pH 8.8)
3. Stacking buffer (1M Tris-HCl (pH 6.8)
4. TEMED
5. 10 per cent ammonium persulphate (1.25g in 10m1 distilled water)
6. 5X running/reservoir buffer (94 gm Glysine, 15 1 gm Tris base, 50 ml 10 per cent SOS. Dissolve the components in 800 ml distilled water. Adjust the volume to 1000 in!). Working concentration is IX
7. Staining solution (1.25g **of** Coomassie Brilliant Blue-R 250 in 1 litre of Methanol : Acetic acid : Water (4:1:5:))
8. Destaining solution (Methanol:acetic acid: Water in the ratio of 4:1:5)

Procedure

Cassette Preparation

1. Clean the glass plates (both notched and rectangular) with methanol or rectified spirit soaked cotton and swab with Whatman paper.
2. Assemble the glass plates with I mm spacers and clamp stand on the level table.
3. Add 1 per cent agar in IX running buffer to the bottom of the cassettes for sealing.
4. After 10 minutes remove the solidified after traces deposit along the internal space of the gel cassette with No.1 Whatman filter paper strips.

Gel Casting and Loading of Protein Sample

1. Prepare the separating gel solution according to required volume (given in the following table).
2. Pour the separating gel solution with 20 ml syringe in the cassette.
3. Overlay the separating gel with water-saturated n-butanol and leave for 1 hour for complete polymerization.
4. After 1 hour remove the n-butanol layer and wash the cassette with distilled water twice.
5. Prepare the stacking gel (as shown in the following table) and pour in to cassette and insert the comb for well formation.
6. After 1 hour moist the stacking gel with distilled water, remove the comb carefully and clean the well with X-film strips.
7. Assemble the gel onto the gel apparatus and fill the upper and lower tank with 1X running buffer and load 50 – 100 i_tg of protein sample in the wells.
8. Connect the power supply to gel unit, initially run at low voltage and after crossing the stacking gel increase the voltage and run at 4°C till dye front reaches the agar.

Staining

1. After complete running switch off the power supply and remove the gel cassette from the gel unit,
2. Disassemble the gel cassette with the help of scalpel and moist the gel with distilled water, remove the stacking gel and agar layer and transfer the separating gel in to staining tray with CBB-R 250 for 4 hrs to 0/N.
3. Destain in 10 per cent acetic acid till background gets cleared and protein bands are clearly visible.

13.4. Western Blotting using Semidry Apparatus

Requirements

(a) As that of in protocol 13.3

(b) Transfer buffer (3.0 gm Tris base. 14.4 gm glycine. 0.1 per cent SDS. Add 10 per cent methanol prior to use)

(c) Ponceau S stain (0.1 gm Ponceau S in 1 per cent acetic acid)

(d) Fat free milk powder

(e) 10X PBS

(f) PBS-T (1X PBS and 0.1 per cent Tween-20, IX PBS and 0.2 per cent Tween-20. IX. - PBS and 0.3 per cent Tween-20)

(g) 2,3-diamino benzidine tetra hydrochloride (I mg/m1 in 1X PBS and 0.014 per cent hydrogen peroxide) OR ECL detection system

Procedure

1. Prepare the SDS-PAGE gel and load the sample and run the sample 'according to the protocol 13.3.1 and 13.3.2.
2. Trim the unwanted portions of the gel. Make a cut on one Coiner of the gel as identification mark and transfer the gel to transfer buffer.
3. Cut 30 pieces of Whatman No. 1 filter paper and soak the filters in the transfer buffer atleast for half an hour in the transfer buffer.
4. Cut nitrocellulose or PVDF membrane exactly to the size of the gel.
5. If nitrocellulose membrane is used, prewet the filter in distilled water and transfer to transfer buffer. If PVDF membrane is used, prewet in methanol and soak in water and transfer to transfer buffer, Make a cut to a corner for identification as made for the gel.
6. Place 10 pieces of filter on the anodic plate of the semidry transfer apparatus. Place the gel over the filters. Then place the membrane over the gel by matching the marks of identification *i.e.*, cut surfaces. Role a glass pipette over the membrane to remove any air bubble between gel and membrane.
7. Place the remaining filters over the membrane.
8. Place the cathodic plate and connect to the'power supply. Run at 100 mA for 2 – 4 hours at 4°C.
9. Remove the membrane and air dry.
10. Stain the membrane with Ponceau S for 2 – 3 mins and destain with distilled water. Take the photograph of the membrane for further reference. One can store the membrane at 4°C till further use.
11. The non specific proteins are blocked by using 5 per cent fat free milk in 1X PBS for 2 hours at room temperature.
12. Wash the membrane in 0.1 per cent PBS-T for 15 min twice.
13. Incubate the membrane in 1:1000 dilution of primary antibody made in blocking_ buffer.
14. Wash the membrane in 0.1 per cent PBS-T for 15 min twice.
15. Incubate the membrane in 1:1000 dilution of the secondary antibody made in blocking buffer. Wash the membrane in 0.2 per cent PBS-T for 15 min once and 0.3 per cent PBS-i' for 15 min once.
16. The signal can he detected with 1 malml. of DAB with an enhancer such as cadmium chloride or nickel chloride or using a chemiluminiscent detection system such as FCL Plus chemiluminescent detection system (Arnersahm Biosciences).

— Section B —

Bioinformatics

14

Bioinformatics: Concepts, Tools and Techniques

Bioinformatics is defined as Conceptualizing biology in terms of molecules (as all interaction in body are taking place as molecular/micro level) and applying informatics techniques (Applied mathematics, computer science, IT, Information Science and statistics to understand and organize the information associated with these molecules on a large scale. It is also called as MIS for molecular Biology

Bioinformatics is a frontier and interdisciplinary area which involves application of mathematics, computer science, and information technology *etc.* in biological sciences.

It involves use of it tools to describe, analyze, simulate and predict the dynamics of various biological processes The ultimate goal of bioinformatics is to enable the discovery of new biological insights as well as to create a global perspective from which unifying principles in biology can be discerned.

During the last ten years, molecular biology has witnessed an information.

Revolution as a result both of the development rapid DNA sequencing techniques and of the corresponding progress in computer based technologies, which are allowing us to cope with this information deluge in increasingly efficient ways.

In broad sense, the term Bioinformatics can be considered to mean information technology applied to the management and analysis of biological data.

The science of bioinformatics or computational biology is increasingly being used to improve the quality of life as we know it.

Bioinformatics has developed out of the need to understand the code of life. DNA. Massive DNA is the basic molecule of life. It codes for genes, which code for proteins, which determine the biological makeup of humans or any living organism. If there are variation(s) or error(s) in the genomic DNA, they ultimately define the like. Hood of developing diseases or resistance to some disorders.

The ultimate goal of bioinformatics is to uncover the wealth of biological information hidden in the mass of sequences, structure, literature and other biological data and obtain a clear insight into the fundamental biology of organisms and use this information to enhance the standard of life for mankind.

It is being used now and in the foresee able future in the areas of molecular medicine to help produce better and more customized medicines to prevent or cure diseases, it has environmental benefits in identifying waste cleanup bacteria and in agriculture it can be used for producing high yield low maintenance crops. These are just a few of the many benefits bioinformaties will help develop.

Fundamental Issues

How we describe, analyze, simulate and predict the dynamics of various biological processes using IT tools.

Complexity: Due to massive amount of data obtained through numerous Biological experiments.

Managing and interpreting Biological data:

EMBL database: 17.807.926,047 nucleotides

Total entries: 15,851,373

The Challenges of Bioinformatics: Today, Tommorrow and the Near Future

The Three E's

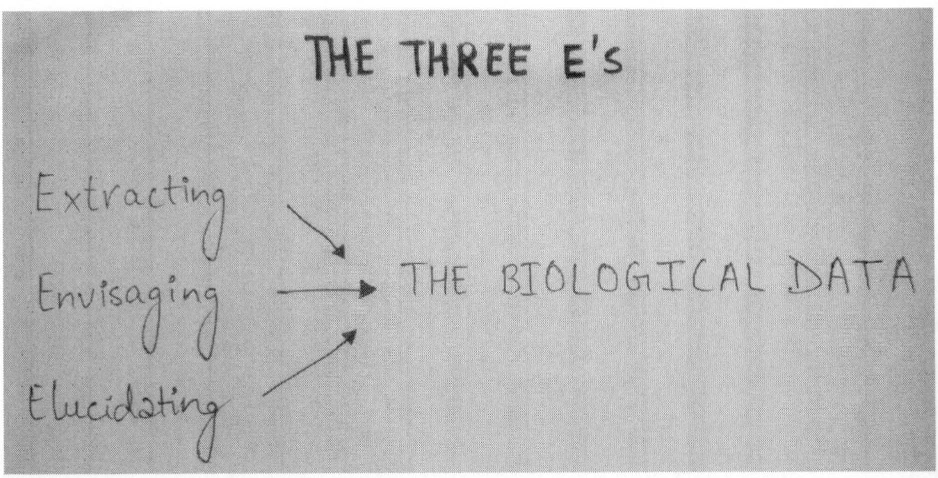

With the recent advances in recombinant DNA technology, cloning *etc.*eteras more and more information owing to cDNA are obtained. And the need of the hour is how this data can be stored and what best of the data could be used and retrieved. For this. Bioinformatics or computational biology has been the solution and front runners in this field. As more and more data is deposited, more and more strategies are to be applied for pragmatic use of computational techniques in biotechnology. Though there are many ways of defining bioinformatics. It is aptly defined taking into consideration the following three:

1. To store the data
2. To maintain the data
3. To retrieve the data

Central Dogma

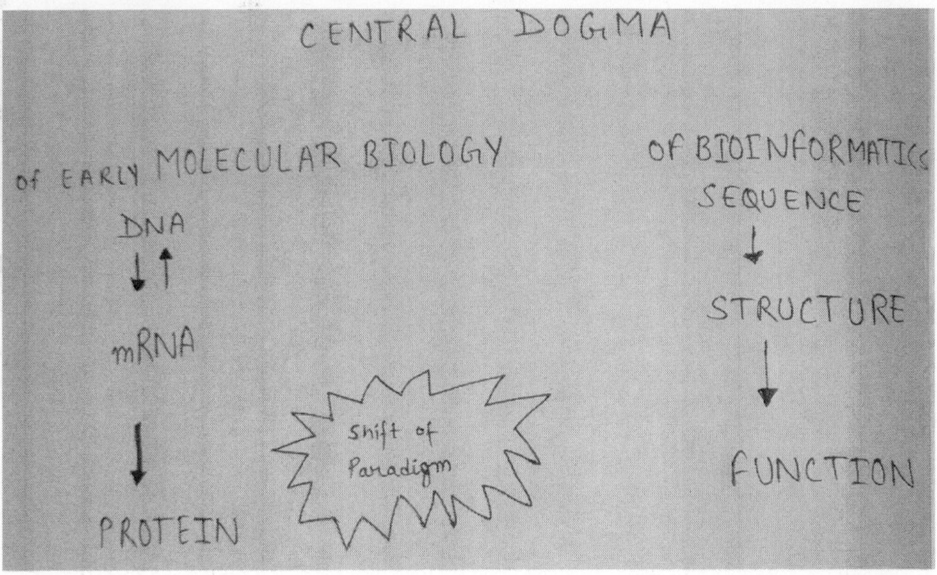

15

Importance of Bioinformatics

In a field which has been dominated by structural biology for the last 2-3 decades, we are now witnessing a dramatic change of focus towards sequence analysis, spurred on by the advent of the genome projects and the resultant sequence/ structure deficit. The prime challenge of bioinformatics is the rationalization of the mass of sequences information, with a view not only to deriving more efficient means of data storage, but also to designing more incisive tools. The imperative that drives this analytical process is the need to convert sequence information into biochemical and biophysical knowledge; for deciphering the structural, functional, and evolutionary clues encoded in the language of biological science.

The greatest challenge facing the molecular biology community today is to make sense of the wealth of data that has been produced by the genome-sequencing product. Traditionally molecular biology research was carried out entirely at the experimental laboratory bench but the huge increase in the scale of data being produced in this genomic era has aroused a need to incorporate computers into research process.

Sequence generation and its subsequent storage interpretation analysis are entirely computer dependent task. However the molecular biology of an organism is very complex issue with research being carried out at different levels including the genome, proteome, transcriptome and metabalome levels following on from the explosion in volume of genomic data similar increase in data been observed respectively in the foeld of proteonomics, transcriptpmics and metabalomics.

The first challenge facing the bioinformatics today is the intelligent and efficient storage of mass of data. It is then their responsibility to provide easy and reliable access to data. The data itself is meaningless before analysis and the sheer volume make it impossible for even a trained biologist to begin to interrupt it manually. Therefore incisive computer must be developed to allow the extraction of the meaningful biological information.

There are three central biological processes bioinformatics tools must be developed:

1. DNA sequence determines protein sequence
2. Protein sequence determines protein structure
3. Protein structure determines protein function

The integration of information learned about these keys biological processes should allow us to achieve the long-term goal of the complete understanding of the biology of organism biological databases.

Biological database are archives of consistent data that are stored in a uniform and efficient manner. These database contain data from a broad spectrum of molecular biology areas. Primary or archived database contain information and annotation of DNA and protein sequences. DNA and protein structures and DNA and protein expression profiles.

Secondary and derived databases are so called because they contain the result of analysis on the primary recourses including information on sequence, patterns or motifs, variants and mutation and evolutionary relationships. Information from the literature is contained in bibliographic databases such as:

Medline (http//www.ebi.ac.uk/database/MEDLINE/mediline.html)

It is essential that these databases are easily accessible and that an Institutive query system is provided to allow researchers to obtain very specific information on a biological subject. The data should be provided in clear consistent manner with some visualization tools to add biological interpretations.

Specific databases for a particular subject have been set up for example.

EMBL database: http://www.ebi.ac.uk/embl/index.html

Biological Applications

Once all the biological data is it stored consistently and is easily available to the scientific community the requirement is then to provide methods of extracting the meaningful information from the mass of data: bioinformatics tools are software programs that are designed to carry out this analysis step.

Factors that must be taken into consideration when designing these tools are:

1. The end user (the biologist) may not be a frequent user of computer technology.
2. These software tools must be made available over the Internet given the global distribution of the scientific research community.

The EBI provides a wide range of biological data analysis tools that fall into the following four major categories;

1. Homology and similarity tools
2. Protein function analysis
3. Structural analysis
4. Sequence analysis

Homology and similarity tools homologous sequences are sequences that are related by divergence from a common ancestor. Thus, the degree of similarity between two sequences can be measured while their homology is a case of being either true or false.

This set of tools can be used to identify similarities between novel query sequences of unknown structure and functions have been elucidated.

Protein Function Analysis

This group of programs allows you to compare your protein sequence to the secondary (or derived) protein databases that contain information on motifs, signatures and protein domains. Highly significant hits against these different pattern databases allow you to approximate the biochemical function of your query protein.

Structure Analysis

This set of tools allows you to compare structures with the known structure databases. The function of protein is more directly a consequence of its structure rather than its sequence with structural homologous tending to share functions. The determination of a protein's 2D/3D structure is crucial in the study of its function.

Sequence Analysis

This set of tools allows you to carry out further, more detailed analysis on your query sequence including evolutionary analysis identification of mutations, hydropathy regions, CpG islands and compositional biases. The identification of these and other biological properties are all clues that aid the search to elucidate the specific function of your sequence.

Bioinformatics and Computational Biology

1. Improve content and utility of databases.
2. Develop better tools for data generation, capture and annotation.
3. Develop and improve tools and databases for comprehensive functional studies.
4. Develop and improve tools for representing and analyzing sequence similarity and variation.
5. Create mechanism to support effective approaches for producing robust, exportable software that can be widely shared.

Bioinformatics Tools

BLAST

BLAST stands for Basic Local Alignment Search Tool and is the most widely used application for rapid screening of large sequence databases. BLAST implementation has been developed and distributed by the National Centre for Biotechnology Information (NCBI) that has become of a "Gold standard" for similarity assessment.

Different Flavors of BLAST

BLASTN – Compares a nucleotide sequence with a nucleotide database.

BLASTP – Compares a protein sequence with a protein database.

T BLASTN – Compares a protein sequence with a nucleotide database.

BLASTX – Translate DNA to protein and then compare with protein database.

TBLASTX – Translate DNA to protein and compare with translated nucleotide database.

BLAST2.0 - Recent version, inserts gaps to optimize the alignment.

Multiple Sequence Alignment

The goal of multiple sequence alignment is to generate a concise, information-rich summary of sequence data in order to inform decision-making on the relatedness of Sequence to a gene family

The Clustal W tool is used for the multiple sequence alignment. Sometime, indeed. Multiple alignments may be used to express the dissimilarity b/w a set of sequences. Alignments should be regarded as models (in both a mathematical and a biological sense) that can be used to test hypothesis Both sequence and structure-based alignments are imperfect models, as neither can reflect all of the evidence.

FASTA

FASTA Format is often required when searching molecular databases, the first line begin with "." and a short description followed by the sequence. Description can be anything that you use choose to write FASTA report from different organizations can be obtained, they are-

Genbank

EMBL

DDBI

DOTPLOT

Comparing sequences is one of the most difficult tasks biologists ever came across. Dot Plots are the simplest means of comparing two sequences. Dodiet is one of the most friendly dot plot programs available on the net. It is a java applet that you can access on the EMBnet Server.

Judging Similarity

The way of interpreting homology to know similarity in human beings, we consider the features, in a similar way, to judge similarity among the sequences; there are attributes for the sequences, which tend to be similar. This is indicated using a score. So if the score is high, we may consider that sequence to be more similar. The homology search is done commonly using programs like BLAST and FASTA. For understanding the concept of sequence alignment, a module called dynamic programming is used uncovering structural and functional information. The main objective of the sequence analysis would be to find sequences in relate to structural and functional needs. However there are other tasks that are to be focused like predicting structure and functions apart from molecular classification.

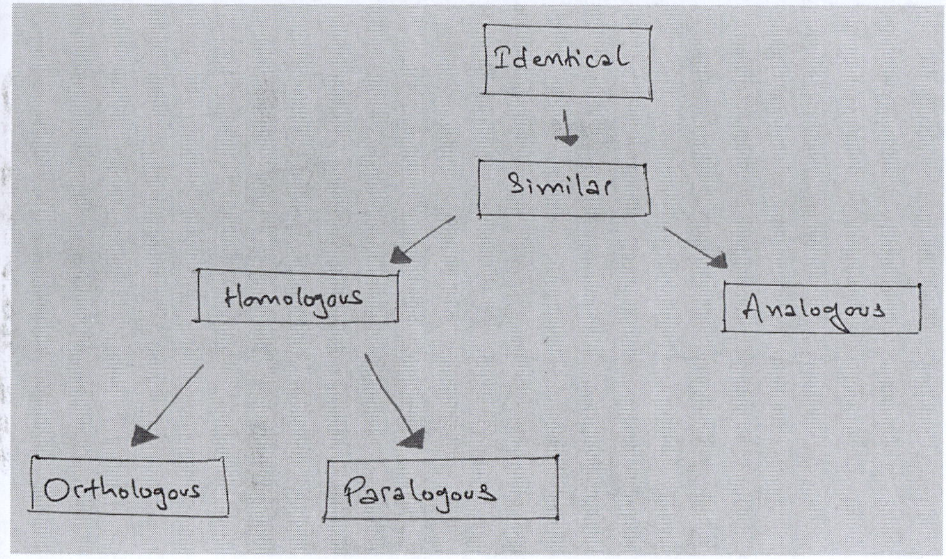

Diagram Showing the Identical Distribution.

There are various problems in similarity searches especially in sequence analysis This so-called computational biology tools would decipher the needs for such purpose in.

BLAST vs. FASTA

BLAST	*FASTA*
Basic Local Alignment Search Tool	No full form, the name given from the way of application, the format/algorithm used.
http://www.ncbi.nlm.nih.gov/BLAST	http:www.ebi.ac.uk/FASTA33/index.html
Works online	Also used for standalone systems
Involves characteristic property of dot matrix and used for high similar sequences	Usually for less similar sequences
Other types of blast like BLASTn, BLASTp, tBLASTn, BLASTx	Other types of FASTA like WU Blast etc.

Remember

- Open Reading Frame (ORF) is a set of long DNA sequences with no interruptions of DNA termination codons. These are all in triplets.
- Today there is many a bioinformatics software which could give information about the satellite DNA sequences repeat sequences, inverted repeats or palindromic repeats, other polymorphic DNA *etc.*etera
- BLAST stands for Basic Local Alignment Search Tool (http://www.ncbi.nlm.nih.gov/BLAST)
- FASTA is also for local alignment and the sequence morphology starts with > symbol (http://www.ebi.ac.uk/fasta)
- Local alignment vs global alignment local alignment is a part of DNA or a protein sequence specific to the query sequence where as a global alignment refers to the whole query sequence.

Scope and Prospects of Bioinformatics

As new DNA sequences keep on propping up, these could be compared with various databases of known sequences. There might be possibility of getting to know the relationships of the sequences, which plays an important role and acts as a clue to get to know of how of the function Today, there are many databases of repute widely used for DNA and protein sequences with robust, iterative and dynamic programming approaches. And in the near future, with advent of fastest computing technologies the data could be accommodated more easily.

Protein Structural Databases

Nearly all proteins have structural similarities with other proteins and, in some of these cases, share a common evolutionary origin. The SCOP database, created by manual inspection and abetted by a battery of automated methods, aims to provide a detailed and comprehensive description of the structural and evolutionary relationships between all proteins whose structure is known. As such, it provides a broad survey of all known protein folds, detailed information about the close relatives of any particular protein, and a framework for future research and classification.

Databases are

- Structural similarity search of SCOP using SSM
- Combinatorial Extension (CE) method for structural comparison
- PALI pairwise and multiple alignments of SCOP families
- SUPFAM structure/sequence relationships
- Structural similarity search of SCOP using 3dSearch
- Structural alignment of SCOP sequences (database + server)
- PINTS-Pattems In Non-homologous Tertiary Structures

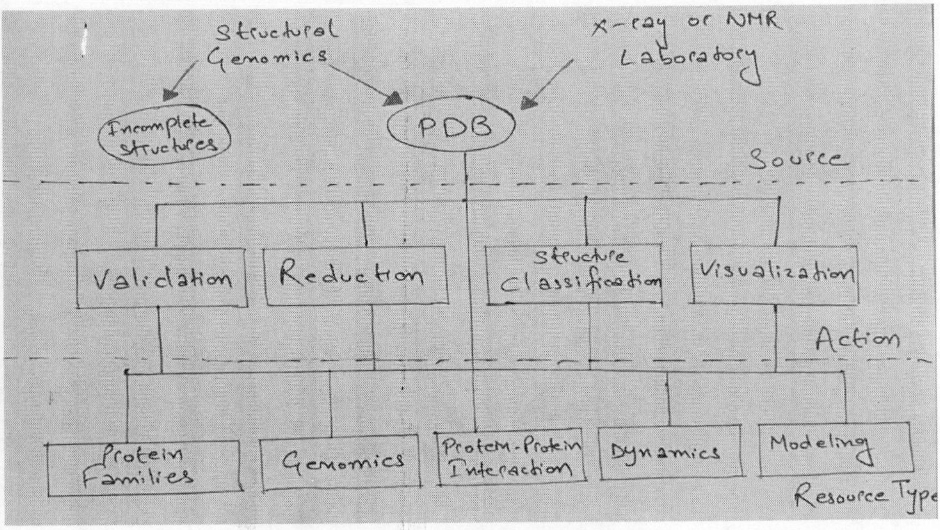

Scheme of Protein Data Bank.

Constructing Proteins-The Genetic Build Up

One DNA strand of each gene acts as a code that specifies the construction of a specific protein. Along the DNA strand of a gene every three nucleotides code for a single amino acid in a protein. A nutation in the DNA of a gene occurs when the sequence of nucleotides in the gene is altered, either by accident during replication of the chromosomes, or by exposure to physical or chemical agents that damage DNA An alteration in the nucleotide sequence of a gene can change the structure of the protein for which it codes Proteins are composed of individual amino acids linked together like the flakes attached with one another. The number of amino acids in a protein differs from one protein to the other Amino acids differ in the type of side chains they contain, which gives each amino acid certain characteristic chemical properties.

Shape of a Protein-Prelude Folding

The shape of a protein gives the characteristic fold to the structure which in turn comprises various shapes like linear, globular and other complex structures. The fold is not distinct and unilateral but there are series of transformations like linear to globular, globular to complex *etc.* The conformation that the protein assumes depends on the order and arrangement of the specific amino acids from which it is constructed. The protein folds into a shape that best accommodates these different compatibilities within the chain, and with the water around it. The shape of a protein determines its function because its shape specifies the other molecules with which it can interact.

A mutation in the gene that codes for a protein can cause structural changes in the protein that affects its shape. In time a change in the protein shape can alter the purpose of the protein, either making it non-functional in terms of its normal role in the cell, or enabling it to acquire a new function that can either be beneficial or destructive to the cell

Using Blast to Compare my Protein Sequence to Other Protein Sequences

BLAST (short for Basic Local Alignment Search Tool-with a name like that, no wonder they shortened it to BLAST) is a great sequence comparison tool that quickly tells us which of the other known proteins out there has a sequence similar to ours We can then use this information for a variety of purposes, including the prediction of protein function. 3-D structure and domain organization, or identification of homologues in other organisms

BLAST comes is us under the auspices of the National Center for Biotechnology Information (NCBI), familiar to us as the hosts for the PubMed bibliographic database:

1. **Point the favorite Internet browser to www.nlm.nih.gov/BLAST**

   The BLAST home page, probably the most frequented bioinformatic web page in the world appears.

2. **Click the Standard Protein-Protein BLAST link**

   The query screen appears

3. **Open the dUTPase FASTA-formatted protein sequence file**

   This is the file that we created on cus PC be using the step in the "Retrieving a li af related protein sequences

4. **Using our browser's menu, copy and paste ONE of the protein sequences into the BLAST search window**

   At this point we need FASTA-formed protein sequence.

5. **Deselect the Do CD- Search hos, but don't change the Choose Database setting.**

   The nit (for nonredundant) default database includes all protein sequence known to man, so we don't have to bother about selecting a specific organism or subset

6. **Click the BLAST! Button.**

7. **Click the Format! Button**

   BLAST now gives us an estimated running time of our query. If the time of the day is right, this shouldn't be more than few seconds.

8. When the result page appears, scroll down the page until we reach a long list of sequence. What we have here ute, the sequence that have a significant similarity with our query sequence, ranked by decreasing the score values.

9. Click one of the underlined scores in the second column from the right. This brings us further down in the page, to see the alignment between our query sequence and the matching ne sequence of the protein corresponding to this score.

In our example, the best match is an identical one. By going down the list, you can see less perfect matches, slowly degrading as the corresponding score decreases and the E value increases. The E-value is the assessment of the statistical significance of the score. E-values close to 1 are warning that the conclusion that we might draw from the alignment is not reliable.

A Multiple Protein Sequence Alignment with ClustalW

Besides running database searches to identify similar protein in a pairwise manner, the second most common bioinformatic tank that biologists like to perform with protein sequence is a multiple alignment. Multiple alignment consists in lining up many similar proteins side by side for the sake of comparison. Multiple alignments are used to

1. Identify sequence position where specific amino acids really matter for the structure integrity or the function of a given protein.
2. Define specific sequence signatures for protein families.
3. Classify sequences and build evolutionary trees.

Do the following to get our feet wet uning by Chuntal W

1. **Point our browser to** http://www.pir.georgetown.edu.

   The PIR some page appears
2. **Under the search and retrieval heading click the ClustalW or Alignment link in the scrall-down menu to display the input form.**

   The input form appears. At this we ned a lew FASTA-formatted protein sequences.
3. **Opm the FASTA-formatted sequence file** alternatively
4. **Copy/paste five of these sequences in the input window**.

   Be careful to copy the definition line and the entire amino acid sequence for each of them
5. **Click the submit button.**

   Within a few seconds, our screen is proudly displaying its first seal one and multiple alignment.

This output illustrates the power of multiple alignments. It clearly delineates a region of high versus low sequence conservation in these related proteins Position

100 per cent conserved or nearly conserved tend to occur in a clustered fashion along sequence defining Various types of functional signatures can be extracted from such a multiple alignment evolutionary relationship between proteins are inferred from phylogenetic trees.

Comparing Two Sequences

Typically, biologists use pairwise comparisons to refine database search results and carry out detailed analysis. It is a common practice to use a pairwise techniques after database search in order to check whether some of the reported matches are really interesting or are just a flash in the plan

Choosing the Right Sequences

The best way to find a sequence homologous to a sequence we already have is to make a database search with a BLAST program. Select out sequences by using following

DNA sequences: At fra 70 per cent identity over more than 100 bases between the hit the query or an E-value lower than a 0.001.

Protein sequence: More than 25 per cent identity over more than 100 amino acids between the hit and the quiry or an E-value lower than 0.0001.

Making a Dot Plot

Dot plots are the simplest means of comparing two sequences. Dot plot techniques are closelt related to sliding window methods-where the window is a portion of our sequence That we compare with other sequences portion of similar size. Dot plots have major advantages. They are simple and don't require any biological hypothesis.

Using Dotlet over the Internet

The most wer-friendly dotlet programs available over the internet. It is a jus applet that we can ances in the EBMnat server. We can download Dole in one click of ampling penting ine browser to www.isree.isbsib.ch/java/dotlet/Dotlet.html.

The output shows Dodet's three windows:

1. **The dot-plot window:** The large window on the left contains our dot plot
2. **The threshold window:** This small window on the right in the one we must to time the sensitivity of your dot plot.
3. **The alignment window:** This is the long narrow window below that shows which pair of residue from the two sequences responds to each dot in the dot plot window.

Anytime we see graphic pattern, we knows we are looking at tandem repeats If this is the case we can use this collection of diagonals to draw three interesting deductions:

1. The number of repeats is equal to the number of diagonals above the main diagonal.
2. The distance between two adjacent diagonals represents the size of the repeats if no spacing falls between the repeats.
3. The shortest diagonal gives us the coordinates of a single repeat unit.

Appendix

Appendix

Robb's Minimal Medium by Fristorm *et al.* (1973)

Chemicals	*Final Concentration*
NaCl	40.00mM
KCl	40.00mM
Glucose	10.00mM
$MgCl.6H_2O$	01.20mM
$MgSO_4.7HO$	01.20mM
KH_2PO_2	00.40mM
NaH_2PO_4	00.40mM
Serine	00.16mM
Valine	00.64mM
Glutamine	04.00mM
Asparagine	00.20mM
Proline	00.32mM
Leucine	00.64mM
Glycine	00.16mM
$CaCl_2$	01.00mM

Dissolve all components in sterilized distilled water and CaCl, is added at last slowly with agitation. Adjust the pH between 72-7.4, filter and store at 4°C

30 per cent Acrylamide

Acrylamide	29.2gm
N, N Bismethyleneacrylamide	00.8gm

Dissolve the contents in 100ml sterilized double distilled water, filter and store in dark Bottle at 4°C.

IM Tris pH 8.8

Tris	12.12gm
Distilled water	80.00ml

Adjust the pH with HCl and make up the volume to 100ml. sterilize and store it at 4°C

IM Tris Cl pH 6.8

Tris	12.12gm
Distilled water	80.00ml

Adjust the pH to 6.8 with HCl and make up the volume to 100ml and sterilize it. Store it at 4°.

10 per cent SDS

SDS	10.0gm
Sterilized distilled water	100.0ml

Dissolve in of and store it at room temperature.

Destaining solution

Acetic acid	20 per cent
Methanol	50 per cent

5X TGS Buffer (Running buffer)

Glycine	94.0gm
Tris base	15.1gm
10 per cent SDS	50.0ml

Dissolve all the components in distilled water and make the volume to 1 litre

1 per cent Agar agar (Qualigen)

Agar agar	1.00gm
Running buffer	100.00ml

Ponceaue S stain

Ponceaue S stain	100mg
1 per cent acetic acid	100ml

Transfer Buffer

Tris	1.5gm
Glycine	7.5gm
Methanol	50.0ml

Dissolve the components in 450 ml of distilled water and store at 4^{0}

Blocking Buffer for Immunoblotting

Skimmed milk powder	5gm
IX PBS	100ml
Tween 20 (sigma)	100µl

Washing Buffer for Immunoblotting

1X PBS	100ml
Tween 20	100µl

10X PBS

$NaH\ PO_4$	0.936gm
Na_2HPO_4	2.492gm
NaCl	15.210gm

Dissolve the contents in 200ml of distilled water and adjust the pill at 7.2.

Washing Buffer for Immunostaining

IX PBS BSA	0.3 per cent
Triton X-100	0.3 per cent

50 per cent GLYCEROL

Glycerol	50ml
Sterilized distilled Water	50ml

Store at 4 degree Celsius.

10 per cent APS

APS	100mg
Distilled water	1ml

Sample Buffer 2X

1M Tris Cl Ph 6.8	00.5ml
10 per cent SDS	02.0ml
Glycerol	01.0ml
Distilled Water	01.5ml
Bromophenol blue	10.0mg

Dissolve the contents and make the aliquots in eppendorf tube. Sample buffers are stored at -20 degree Celsius.

Sample Buffer 1X

Sample buffer2x	50 per cent
Distilled Water	45 per cent
β-mercaptoethanol	05 per cent

10 per cent TCA (Trichloroacetic acid)

TCA	05.0gm
Distilled Water	22.7ml

1N NaOH

NaOH Pellet	400mg
Distilled Water	10ml

0.15M NaCl

NaCl	790mg
Distilled water	100ml

CBB G-250 dye

CBB	50mg
Ethanol 95 per cent	25ml
Phosphoric Acid 85 per cent	50ml
Distilled Water	375ml

0.1 per cent CBB R-250 Stain	**100ml**
Methanol	20ml
Glacial acetic acid Distilled Water	80ml
CBB G- 250	200mg

Blocking Buffer

IX PBS	3.0 per cent

BSA	0.5 per cent
Triton X-100	

0.3 per cent H_2O_2

H_2O_2	30µl
70 per cent methanol	100ml

DAB (working solution)

DAB stock	200µl
IX PBS	770µl
H_2O_2	30µl

4 per cent Paraformaldehyde

Paraformaldehyde	400mg
IX PBS	10ml

Index

S

T

W